T0122203

Intelligent Systems Reference Library

Volume 147

Series editors

Janusz Kacprzyk, Polish Academy of Sciences, Warsaw, Poland
e-mail: kacprzyk@ibspan.waw.pl

Lakhmi C. Jain, University of Canberra, Canberra, Australia;
Bournemouth University, UK;
KES International, UK
e-mail: jainlc2002@yahoo.co.uk; jainlakhmi@gmail.com
URL: http://www.kesinternational.org/organisation.php

The aim of this series is to publish a Reference Library, including novel advances and developments in all aspects of Intelligent Systems in an easily accessible and well structured form. The series includes reference works, handbooks, compendia, textbooks, well-structured monographs, dictionaries, and encyclopedias. It contains well integrated knowledge and current information in the field of Intelligent Systems. The series covers the theory, applications, and design methods of Intelligent Systems. Virtually all disciplines such as engineering, computer science, avionics, business, e-commerce, environment, healthcare, physics and life science are included. The list of topics spans all the areas of modern intelligent systems such as: Ambient intelligence, Computational intelligence, Social intelligence, Computational neuroscience, Artificial life, Virtual society, Cognitive systems, DNA and immunity-based systems, e-Learning and teaching, Human-centred computing and Machine ethics, Intelligent control, Intelligent data analysis, Knowledge-based paradigms, Knowledge management, Intelligent agents, Intelligent decision making, Intelligent network security, Interactive entertainment, Learning paradigms, Recommender systems, Robotics and Mechatronics including human-machine teaming, Self-organizing and adaptive systems, Soft computing including Neural systems, Fuzzy systems, Evolutionary computing and the Fusion of these paradigms, Perception and Vision, Web intelligence and Multimedia.

More information about this series at http://www.springer.com/series/8578

Verónica Bolón-Canedo · Amparo Alonso-Betanzos

Recent Advances in Ensembles for Feature Selection

 Springer

Verónica Bolón-Canedo
Facultad de Informática
Universidade da Coruña
A Coruña
Spain

Amparo Alonso-Betanzos
Facultad de Informática
Universidade da Coruña
A Coruña
Spain

ISSN 1868-4394   ISSN 1868-4408   (electronic)
Intelligent Systems Reference Library
ISBN 978-3-030-07929-1   ISBN 978-3-319-90080-3   (eBook)
https://doi.org/10.1007/978-3-319-90080-3

© Springer International Publishing AG, part of Springer Nature 2018
Softcover re-print of the Hardcover 1st edition 2018
This work is subject to copyright. All rights are reserved by the Publisher, whether the whole or part
of the material is concerned, specifically the rights of translation, reprinting, reuse of illustrations,
recitation, broadcasting, reproduction on microfilms or in any other physical way, and transmission
or information storage and retrieval, electronic adaptation, computer software, or by similar or dissimilar
methodology now known or hereafter developed.
The use of general descriptive names, registered names, trademarks, service marks, etc. in this
publication does not imply, even in the absence of a specific statement, that such names are exempt from
the relevant protective laws and regulations and therefore free for general use.
The publisher, the authors and the editors are safe to assume that the advice and information in this
book are believed to be true and accurate at the date of publication. Neither the publisher nor the
authors or the editors give a warranty, express or implied, with respect to the material contained herein or
for any errors or omissions that may have been made. The publisher remains neutral with regard to
jurisdictional claims in published maps and institutional affiliations.

Printed on acid-free paper

This Springer imprint is published by the registered company Springer International Publishing AG
part of Springer Nature
The registered company address is: Gewerbestrasse 11, 6330 Cham, Switzerland

To our children: Iago, Alberto, Leo and Olivia

Foreword

Ensemble methods are now a cornerstone of modern Machine Learning and Data Science, the "go-to" tool that everyone uses by default, to grab that last 3–4% of predictive accuracy. Feature Selection methods too, are a critical element, throughout the data science pipeline, from exploratory data analysis to predictive model building. There can scarcely be a more generically relevant challenge, than a meaningful synthesis of the two. This is the challenge that the authors here have taken on, with gusto.

Bolón-Canedo and Alonso-Betanzos present a meticulously thorough treatment of literature to date, comparing and contrasting elements practical for applications, and interesting for theoreticians. I was surprised to find several new references I had not found myself, in several years of working on these topics.

The first half of the book presents tutorials, cross referenced to current literature and thinking—this should prove a very useful launch-pad for students wanting to get into the area. The second half presents more advanced topics and issues—from appropriate evaluation protocols (it's really not simple, and certainly not a done deal yet), to still quite open questions (e.g. combination of ranks and the stability of algorithms), through to software tips and tools for practitioners. I particularly like Chapter 10, on emerging challenges. This sort of chapter points the way for new PhDs, providing inspiration and confidence that your research is moving in the right direction.

In summary, Bolón-Canedo and Alonso-Betanzos have authored an eloquent and authoritative treatment of this important area—something I will be recommending to my students and colleagues as essential reference material.

University of Manchester   Prof. Gavin Brown
2018

Preface

Classically, machine learning methods have used a single learning model to solve a given problem. However, the technique of using multiple prediction models for solving the same problem, known as ensemble learning, has proven its effectiveness over the last few years. The idea builds on the assumption that combining the output of multiple experts is better than the output of any single expert. Classifier ensembles have flourished into a prolific discipline; in fact, there is a series of workshops on Multiple Classifier Systems (MCSs) run since 2000 by Fabio Roli and Josef Kittler.

However, ensemble learning can be also thought as means of improving other machine learning disciplines such as feature selection, which has not received yet the same amount of attention. There exists a vast body of feature selection methods in the literature, including filters based on distinct metrics (e.g. entropy, probability distributions or information theory) and embedded and wrapper methods using different induction algorithms. The proliferation of feature selection algorithms, however, has not brought about a general methodology that allows for intelligent selection from existing algorithms. In order to make a correct choice, a user not only needs to know the domain well but also is expected to understand technical details of available algorithms.

Ensemble feature selection can be a solution for the aforementioned problem since, by combining the output of several feature selectors, the performance can be usually improved and the user is released from having to choose a single method. This book aims at offering a general and comprehensive overview of ensemble learning in the field of feature selection.

Ensembles for feature selection can be classified into homogeneous (the same base feature selector) and heterogeneous (different feature selectors). Moreover, it is necessary to combine the partial outputs that can be either in the form of subsets of features or in the form of rankings of features. This book stresses the gap with standard ensemble learning and its application to feature selection, showing the particular issues that researchers have to deal with. Specifically, it reviews different techniques for combination of partial results, measures of diversity and evaluation of the ensemble performance. Finally, this book also shows examples of problems in which ensembles for feature selection have applied in a successful way and introduces the new challenges and possibilities that researchers must acknowledge, especially since the advent of Big Data.

The target audience of this book comprises anyone interested in the field of ensembles for feature selection. Researchers could take advantage of this extensive review on recent advances on the field and gather new ideas from the emerging challenges described. Practitioners in industry should find new directions and opportunities from the topics covered. Finally, we hope our readers enjoy reading this book as much as we enjoyed writing it.

We are thankful to all our collaborators, who helped with some of the research involved in this book. We would also like to acknowledge our families and friends for their invaluable support, and not only during this writing process.

A Coruña, Spain   Verónica Bolón-Canedo
March 2018   Amparo Alonso-Betanzos

Contents

Chapter 1
Basic Concepts

Abstract In the new era of Big Data, the analysis of data is more important than ever, in order to extract useful information. Feature selection is one of the most popular preprocessing techniques used by machine learning researchers, aiming to find the relevant features of a problem. Since the best feature selection method does not exist, a possible approach is to use an ensemble of feature selection methods, which is the focus of this book. But, before diving into the specific aspects to consider when building an ensemble of feature selectors, in this chapter we will go back to the basics in an attempt to provide the reader with basic concepts such as the definition of a dataset, feature and class (Sect. 1.1). Then, Sect. 1.2 comments on measures to evaluate the performance of a classifier, whilst in Sect. 1.3 different approaches to divide the training set are discussed. Finally, Sect. 1.4 gives some recommendations on statistical tests adequate to compare several models and in Sect. 1.5 the reader can find some database repositories.

This book is devoted to explore the recent advances in ensemble feature selection. Feature selection is the process of selecting the relevant features and discarding the irrelevant ones but, since the best feature selection method does not exist, a possible solution is to use an ensemble of multiple methods. But, before entering into specific details when dealing with ensemble feature selection, this chapter will start by defining basic concepts that will be necessary to understand the more advanced issues that will be discussed throughout this book.

1.1 What Is a Dataset, Feature and Class?

This introductory chapter starts by defining a cornerstone in the field of Data Analysis: the data itself. In the last few years, human society collects and stores vast amounts of information about every subject imaginable, leading to the appearance of the term *Big Data*. More than ever, data scientists are now in need, aiming at extracting useful information from a vast pile of row data. But let's start from the beginning... *What is data?*

Data is usually collected by researchers in a form of a dataset. A *dataset* can be defined as a collection of individual data, often called *samples*, *instances* or *patterns*.

© Springer International Publishing AG, part of Springer Nature 2018
V. Bolón-Canedo and A. Alonso-Betanzos, *Recent Advances in Ensembles for Feature Selection*, Intelligent Systems Reference Library 147, https://doi.org/10.1007/978-3-319-90080-3_1

A sample can be seen as information about a particular case, for example about a medical patient. The information about this particular case is given in the form of *features* or *attributes*. A feature might be the sex of the patient, his/her blood pressure or the color of his/her eyes. A feature can be relevant or not, or even redundant with others, but this issue will be explored in depth in Chap. 2.

A specific task in Data Analysis is called classification, which consists of assigning each sample to a specific *class* or category. Typically, samples belonging to the same class have similar features and samples belonging to different classes are dissimilar. A simple example can be seen in Table 1.1, which represents the popular "play tennis" dataset [1].

As can be seen, this toy example represents data for a total of 15 records or samples, and each sample has four different features (outlook, temperature, humidity and windy) which give information that can be useful to know if it is possible to play tennis or not (given that tennis is a sport that is played outside). The last column represents the prediction variable or class (play), which is the desirable outcome of this dataset, in a typical classification scenario. A feature can be discrete (if it takes a finite set of possible values), continuous (if it takes a numerical value) or boolean (if it takes one of two possible values—for example 0 or 1), and in some cases it is necessary to *discretize* the continuous values since some machine learning algorithms can only work with discrete data. In the "play tennis" dataset, features "outlook" and "temperature" are discrete, whilst features "humidity", "windy" and the class "play" are boolean (notice that a boolean feature is a particular case of a discrete feature).

Table 1.1 Play tennis dataset

Outlook	Temperature	Humidity	Windy	Play
sunny	hot	high	false	no
sunny	hot	high	true	no
sunny	hot	high	true	no
overcast	hot	high	false	yes
rainy	mild	high	false	yes
rainy	cool	normal	false	yes
rainy	cool	normal	true	no
overcast	cool	normal	true	yes
sunny	mild	high	false	no
sunny	cool	normal	false	yes
rainy	mild	normal	false	yes
sunny	mild	normal	true	yes
overcast	mild	high	true	yes
overcast	hot	normal	false	yes
rainy	mild	high	true	no

More formally, we can represent a dataset as $\mathbf{X} = \{\mathbf{x_1}, \ldots, \mathbf{x_d}\} \in \mathbb{R}$. The class label is represented as $\mathbf{Y} = \{\mathbf{y_1}, \ldots, \mathbf{y_N}\}$. A typical dataset is organized as a matrix of N rows (samples) by d columns (features), plus an extra column with the class labels:

$$\mathbf{X} = \begin{bmatrix} x_{11}, & x_{12}, & \ldots & x_{1d} \\ x_{21}, & x_{22}, & \ldots & x_{2d} \\ \vdots & & & \\ x_{N1}, & x_{N2}, & \ldots & x_{Nd} \end{bmatrix} \quad \mathbf{Y} = \begin{bmatrix} y_1 \\ y_2 \\ \vdots \\ y_N \end{bmatrix}$$

Notice that the element x_{ji} contains the value for the ith feature of the jth sample.

One of the most popular datasets that can be found in the Pattern Recognition literature is the Iris dataset [2]. This dataset has been used in thousands of publications over the years, and consists of distinguishing among three classes of iris plant (setosa, virginica and versicolor). The dataset has four features which are petal width and length, and sepal width and length and 50 samples of each of the three classes. As can be seen in Fig. 1.1, one of the classes (setosa) is linearly and clearly separable from the other two, while the classes virginica and versicolor are not linearly separable between them. Notice that in Fig. 1.1 we are displaying feature petal width versus petal length, but this situation on separability occurs for each pairwise combination of features.

Having features that are linearly separable leads to perfect classification accuracies, while when the classes are not separable it is possible that the classifiers make some mistakes. This issue will be commented in detail in the next section.

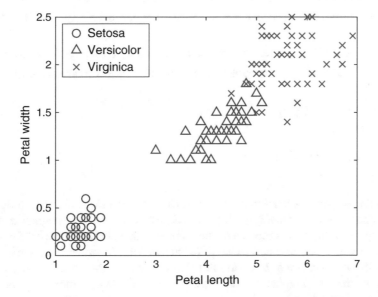

Fig. 1.1 Scatter plot of Iris dataset

1.2   Classification Error/Accuracy

Although this book is focused on feature selection, a typical measure to evaluate the efficiency of the features selected by a feature selection algorithm is to use a classifier afterwards and check if the classification error/accuracy remains acceptable.

Just to recall, the task of a classifier is to predict to which class belongs a particular sample. Therefore, we need measures to evaluate how good this prediction was. A very popular performance measure is the *classification error*, which is the percentage of incorrectly classified instances divided by the total number of instances. Analogously, *classification accuracy* is the percentage of correctly classified instances divided by the total number of samples.

However, looking only at the classification error or accuracy is not a good practice. Suppose that we have a dataset with 100 samples, 95 of them belonging to class A and only five of them belonging to class B. Imagine now that we have two classifiers, C_1 and C_2. The first classifier, C_1, just assigns all the samples to class A, achieving 95% of accuracy, which sounds fairly high. Then, the second classifier, C_2, misclassifies four samples belonging to class A and two samples belonging to class B, obtaining 94% of accuracy. Which classifier is better? Well, the answer depends on the nature of the dataset but, in general, it is better to achieve a trade-off between the performance on the two classes, and so it is necessary to check the classification rates of each class. In a typical binary classification scenario, there are other measures that we can use to evaluate the performance of a classifier, which are represented below. Notice that accuracy and error can be redefined in terms of these new measures.

- *True positive (TP)*: percentage of positive examples correctly classified as so.
- *False positive (FP)*: percentage of negative examples incorrectly classified as positive.
- *True negative (TN)*: percentage of negative examples correctly classified as so.
- *False negative (FN)*: percentage of positive examples incorrectly classified as negative.
- $Sensitivity = \frac{TP}{TP+FN}$.
- $Specificity = \frac{TN}{TN+FP}$.
- $Accuracy = \frac{TN+TP}{TN+TP+FN+FP}$.
- $Error = \frac{FN+FP}{TN+TP+FN+FP}$.

Another way to check how the errors are distributed across the classes (particularly interesting if the dataset has more than two classes) is to construct a *confusion matrix*. An entry a_{ij} of this matrix represents the number of samples that have been assigned to class c_j while their true class was c_i. To calculate the classification accuracy from this matrix it is necessary to divide the sum of the elements in the main diagonal divided by the total number of examples. Using the confusion matrix is very useful because it gives additional information on *where* the errors have occurred.

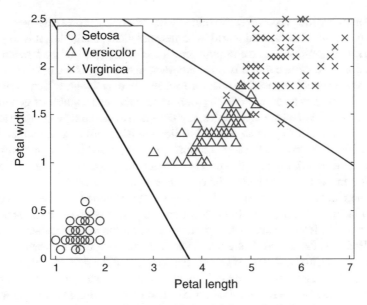

Fig. 1.2 Scatter plot of Iris dataset being classified with a linear discriminant

Table 1.2 Confusion matrix for Iris dataset classified with a linear discriminant

		Predicted		
		Setosa	Versicolor	Virginica
Actual	Setosa	50	0	0
	Versicolor	0	48	2
	Virginica	0	4	46

For example, suppose that we have classified the Iris dataset with a linear discriminant [2]. In Fig. 1.2 we can see that, as expected, the class setosa is correctly classified but there are some errors in the classification of the other two classes. In particular, the confusion matrix presented in Table 1.2 gives us more explicit information about the errors.

1.3 Training and Testing

In the previous section, we have seen, as an example, how the Iris data was classified. But what happens when a new sample arrives? This is the essence of classification, being able to classify *new* examples for which the class label is not known, and in this way test our classification model.

Ideally, one would use all the labeled examples available to train a classifier, making it possible that it can learn the particularities of the data and the relationship between the feature values and the corresponding class. Then, as new unlabeled examples would come, our trained classifier makes a prediction but, how can we know if our classifier was correctly trained with data being representative enough of the full population? In the real world, with new unlabeled examples, it is impossible to answer this question. So, a common practice is to *save* part of the labeled data to act as the *test* set. Notice that it is very important that testing is done on *unseen* data. An important aspect we need to take into account is *overfitting*, which might occur when the learning is so adjusted to the training data that is incapable of generalize to unseen test data. Therefore, in practice, it is common to use some technique to lessen the amount of overfitting, such as cross-validation (that will be commented later in this section), regularization, early stopping, pruning, etc. All these techniques are based on either explicitly penalize overly complex models or to test the ability of the model to generalize by evaluating its performance in unseen data.

All the parameters involved in learning should be tuned on the training data, and this includes feature selection. A commonly found mistake in the specialized literature is that feature selection is performed on all the available data, discarding the irrelevant features, and then continue with the training of the classifier dividing data into training and test sets to evaluate the accuracy of the selected features. This is incorrect, since feature selection (and any other type of learning or parameter tuning that is performed on data) should be done only on the training set, leaving a test set to evaluate the performance.

There are some benchmark datasets that come originally divided into training and test sets. For example, the KDD (Knowledge Discovery and Data Mining Tools Conference) Cup 99 dataset is a benchmark for intrusion detection systems. Separate training and test sets were released, with the particularities that the percentage of the different classes (normal connection and several types of attacks) varies significantly from training to test, as well as the fact that in the test set there are new attacks that are not present in the training set [3].

In other cases, researchers need to keep part of the available data as test set. There are several training/testing protocols that can be done, the most popular ones are following described:

- k-**Fold Cross-validation**. This is one of the most famous validation techniques [4]. The data (D) is partitioned into k nonoverlapping subsets $D_1, \ldots, D_k$ of roughly equal size. The learner is trained on $k - 1$ of these subsets combined together and then applied to the remaining subset to obtain an estimate of the prediction error. This process is repeated in turn for each of the k subsets, and the cross-validation error is given by the average of the k estimates of the prediction error thus obtained. In the case of feature selection, note that with this method there will be k subsets of selected features. A common practice is to merge the k different subsets (either by union or by intersection) or to keep the subset obtained in the fold with the best classification result.
- **Leave-One-Out Cross-validation**. This is a variant of k-fold cross validation where k is the number of samples [4]. A single observation is left out each time.
- **Bootstrap**. This is a general resampling strategy [5]. A *bootstrap sample* consists of n samples equally likely to be drawn, with replacement, from the original data. Therefore, some of the samples will appear multiple times, whereas others will

not appear at all. The learner is designed on the bootstrap sample and tested on the left-out data points. The error is approximated by a sample mean based on independent replicates (usually between 25 and 200). Some famous variants of this method exist, such as *balanced bootstrap* or *0.632 bootstrap* [6]. As in the previous methods, there will be as many subsets of features as repetitions of the method.

• **Holdout Validation**. This technique consists of randomly splitting the available data into a disjoint pair training test [4]. A common partition is to use 2/3 for training and 1/3 for testing. The learner is designed based on the training data and the estimated error rate is the proportion of errors observed in the test data. This approach is usually employed when some of the datasets in a study come originally divided into training and test sets whilst others do not. In contrast to other validation techniques, a unique set of selected features is obtained.

The choice of one or another method is not trivial, and it usually depends on the size of the data we have. For example, if only a hundred of samples are available (as usually happens with microarray data), choosing a 2/3-1/3 hold validation might not be a good idea, since the training data might not be enough to avoid overfitting [7]. On the contrary, when the data is really large (as it happens nowadays since the advent of Big Data), using a 10-fold cross validation or leave-one-out can result in an excessively time-consuming process, so people tend to go back to the old good hold-out method [8]. Moreover, using a scheme that produces multiple training and testing pairs, there is the open question of which of the models built during the training process should be used in the end. For example, imagine that you have used a 10-fold cross validation to perform feature selection and evaluated the performance of the selected features in terms of classification accuracy. You end up with ten — possibly— different subsets of features, and then…which one would you use as your *final* set of relevant features? There is not a perfect solution to this problem, some approaches consist of choosing the one which obtains the highest classification accuracy, while others employ the union or intersection between all the ten subsets of features.

1.4 Comparison of Models: Statistical Tests

When presenting a new feature selection or classification method, it is necessary to compare it with previous state-of-the-art approaches, to demonstrate if the method is sound. For example, if one wants to demonstrate that applying feature selection is useful in a particular domain, a common practice is to compare the classification performance with and without feature selection, expecting that feature selection — at least— maintains the original performance but using less features. Therefore, to know if the differences between models are important, statistical tests are usually employed.

When comparing models, there is a set of good practices that is advisable to follow, based on those given by Kuncheva [8]:

- Choose carefully the training/test procedure (see previous section) before starting the experiments. When you publish your work, give enough details so the experiments are clear and can be reproducible.
- Make sure that all the models (either if we are comparing feature selection methods or classifiers) use all the information possible, and of course they employ the same data for training, and then for testing. For example, it is not fair to perform different 10-fold cross validations for different models, because the random division of the data may favor one or another method. The correct way to do this is to divide the data into folds and at each iteration train the different models on the corresponding training data.
- Make sure that the data reserved for testing was not used before in any training stage.
- When possible, perform statistical tests. It is better for the reader to know if the differences in performance between models are *statistically significant* or not.

There are several statistical tests available in the specialized literature; in the following we will describe the most adequate ones for a particular situation based on the recommendations given by Demšar [9].

1.4.1 Two Models and a Single Dataset

Suppose that you have a fixed, single dataset and two algorithms (for example, the same classifier with and without feature selection as a previous step). If we want to have some repetitions to be able to perform statistical tests, it is necessary to repeatedly split the data into training and testing set, and induce our models. For example, a typical choice might be a 10-fold cross validation. Unfortunately, under this situation, it is not possible to apply the classical Student's t-test for paired samples, since this method assumes that the samples need to be independent, and in a cross-validation they are not (two training sets in a 10-fold cross-validation share 80/90% of data instances). To solve this, there are several options:

- The corrected t-test presented by Nadeau and Bengio [10] which corrects the bias presenting a new way to compute the variance.
- The McNemar's test.
- Dietterich [11] proposed to perform a 5×2 cross-validation. In each 2-fold cross-validation, different data is used for training and testing, so we can assume that the variances are unbiased. Since it is computed in a really small sample (2), Dietterich proposed to repeat this process five times.

1.4.2 Two Models and Multiple Dataset

Given that we have access to data repositories such as the UCI Machine Learning Repository, the tendency is to use several datasets to demonstrate that our new method is better than other, for example that using feature selection before classifying is better than *just* classifying. The prevalent approach some years ago to compare two models was to count wins and losses. However, how can we know if an algorithm really *wins*? If our model A wins in 15 datasets and loses only in two, we can say it but, what if they were 10:7? Notice that our samples, in this case, are the number of datasets tested, so this is a really small sample size, making it difficult to draw meaningful conclusions.

Demšar discourages us to use sign tests, as they discard too much information. They only take into account the signs (of differences) but they do not consider the margins by which each model wins. So, in this situation, Demšar proposes the use of Wilcoxon signed rank test [12].

1.4.3 Multiple Models and Multiple Dataset

Another typical scenario is when you have multiple feature selection algorithms which you want to apply before classifying and you want to know which one is the best. According to Demšar, repeating the Wilcoxon test for all pairs is not a good idea, since it is something that you should avoid in significant testing, specially because your sample size is the number of datasets and it is not large enough. Thus, he suggested the use of the Friedman test [13, 14]. This test only tells if the performances of your models differ, so you need a post-hoc test. There are two possible situations: you either compare multiple algorithms, or you compare your novel method (control case) with several existing algorithms. In the first case, you have $k(k + 1)/2$ comparisons (being k the number of models) and Demšar suggested the use of the pairwise Nemenyi case [15]. In the second case, you test $k - 1$ hypotheses (yours vs. every other) and Demšar suggested the use of the Bonferroni–Dunn test [16].

1.5 Data Repositories

Nowadays, there are several data repositories with benchmark datasets in which researchers can find a diverse set of databases to test their novel methods. The most popular ones are listed below:

- *The UC Irvine Machine Learning Repository* (UCI), from University of California, Irvine:
  http://archive.ics.uci.edu/ml/

- *UCI KDD Archive*, from University of California, Irvine:
  http://kdd.ics.uci.edu
- *LIBSVM Database*:
  http://www.csie.ntu.edu.tw/~cjlin/libsvmtools/datasets/
- *Public Data Sets*, from Amazon Web Services:
  http://aws.amazon.com/datasets
- *The Datahub*:
  http://datahub.io/dataset
- *Kaggle datasets*:
  https://www.kaggle.com/datasets

There also exist specialized repositories, for example for microarrays (with the particularity of having much more features than samples) or images.

- *ImageNet*, the most popular collection of public images:
  https://www.kaggle.com/datasets
- *ArrayExpress*, microarray datasets from the European Bioinformatics Institute:
  http://www.ebi.ac.uk/arrayexpress/
- *Gene Expression Omnibus*, microarray datasets from the National Institutes of Health:
  http://www.ncbi.nlm.nih.gov/geo/
- *The Cancer Genome Atlas (TCGA)*, microarray datasets from both the National Cancer Institute and the National Human Genome Research Institute:
  https://cancergenome.nih.gov/
- *Cancer Program Data Sets*, microarray datasets from the Broad Institute:
  http://www.broadinstitute.org/cgi-bin/cancer/datasets.cgi
- *Gene Expression Model Selector*, microarray datasets from Vanderbilt University:
  http://www.gems-system.org
- *Gene Expression Project*, microarray datasets from Princeton University:
  http://genomics-pubs.princeton.edu/oncology/

1.6 Summary

Feature selection is one of the most popular preprocessing techniques, which consists of selecting the relevant features and discarding the irrelevant and redundant ones. Researchers agree that the best feature selection method does not exist, so a good option might be to combine the outcomes of different selectors, which is known as ensemble feature selection. Before exploring in detail this approach, which will be exhaustively tackled throughout this book, this chapter describes some basic concepts that are necessary to know for any machine learning researcher.

We have described basic concepts such as a dataset, a feature and a class, and we have also payed attention to more delicate issues such as the correct choice of an evaluation system or a statistical test. Moreover, we also provide some repositories from which popular benchmark datasets can be downloaded.

References

1. Quinlan, J.R.: Induction of decision trees. Mach. Learn. **1**(1), 81–106 (1986)
2. Fisher, R.A.: The use of multiple measurements in taxonomic problems. Ann. Eugen. **7**(2), 179–188 (1936)
3. Bolón-Canedo, V., Sanchez-Maroño, N., Alonso-Betanzos, A.: Feature selection and classification in multiple class datasets: an application to KDD Cup 99 dataset. Expert Syst. Appl. **38**(5), 5947–5957 (2011)
4. Bramer, M.: Principles of Data Mining. Springer, Berlin (2007)
5. Efron, B.: Bootstrap methods: another look at the jackknife. Ann. Stat. **7**(1), 1–26 (1979)
6. Efron, B., Tibshirani, R.: An Introduction to the Bootstrap, vol. 57. Chapman & Hall/CRC, Boca Raton (1993)
7. Bolón-Canedo, V., Sanchez-Maroño, N., Alonso-Betanzos, A., Benítez, J.M., Herrera, F.: A review of microarray datasets and applied feature selection methods. Inf. Sci. **282**, 111–135 (2014)
8. Kuncheva, L.I.: Combining Pattern Classifiers: Methods and Algorithms. Wiley, New Jersey (2004)
9. Demar, J.: Statistical comparisons of classifiers over multiple data sets. J. Mach. Learn. Res. **7**(1), 1–30 (2006)
10. Nadeau, C., Bengio, Y.: Inference for the generalization error. Advances in neural information processing systems, pp. 307–313 (2000)
11. Dietterich, T.G.: Approximate statistical tests for comparing supervised classification learning algorithms. Neural Comput. **10**, 1895–1924 (1998)
12. Wilcoxon, F.: Individual comparisons by ranking methods. Biometrics **1**, 80–83 (1945)
13. Friedman, M.: The use of ranks to avoid the assumption of normality implicit in the analysis of variance. J. Am. Stat. Assoc. **32**, 675–701 (1937)
14. Friedman, M.: A comparison of alternative tests of significance for the problem of m rankings. Ann. Math. Stat. **11**, 86–92 (1940)
15. Nemenyi, P.B.: Distribution-free multiple comparisons, PhD thesis, Princeton University (1963)
16. Dunn, O.J.: Multiple comparisons among means. J. Am. Stat. Assoc. **56**, 52–64 (1961)

Chapter 2
Feature Selection

Abstract The advent of Big Data, and specially the advent of datasets with high dimensionality, has brought an important necessity to identify the relevant features of the data. In this scenario, the importance of feature selection is beyond doubt and different methods have been developed, although researchers do not agree on which one is the best method for any given setting. This chapter provides the reader with the foundations about feature selection (see Sect. 2.1) as well as a description of the state-of-the-art feature selection methods (Sect. 2.2). Then, these methods will be analyzed on several synthetic datasets (Sect. 2.3) trying to draw conclusions about their performance when dealing with a crescent number of irrelevant features, noise in the data, redundancy and interaction between attributes, as well as a small ratio between number of samples and number of features. Finally, in Sect. 2.4, some state-of-the-art methods will be analyzed to study their scalability, i.e. the impact of an increase in the training set on the computational performance of an algorithm in terms of accuracy, training time and stability.

Ensemble learning is typically applied to classification problems. However, there are ensembles focused on other machine learning tasks, such as feature selection, which is the focus of this book and will be thoughtfully explained in this chapter.

Among machine learning researchers, it is common to have to deal with datasets containing a huge number of features, which derives in an interesting challenge because classical machine learning methods are not able to efficiently deal with such number of input features. As a result, it is typical to apply a preprocessing step to remove irrelevant features and reduce the dimensionality of the problem at hand.

Dimensionality reduction techniques are usually divided into two groups: *feature extraction* and *feature selection*. Feature extraction consists in combining the original features and obtaining a dataset with a reduced number of new features which are a transformation of the original ones (see Fig. 2.1a). One of the most popular methods

---

Part of the content of this chapter was previously published in *Knowledge and Information Systems* (https://doi.org/10.1007/s10115-012-0487-8 and https://doi.org/10.1007/s10115-017-1140-3).

© Springer International Publishing AG, part of Springer Nature 2018
V. Bolón-Canedo and A. Alonso-Betanzos, *Recent Advances in Ensembles for Feature Selection*, Intelligent Systems Reference Library 147,
https://doi.org/10.1007/978-3-319-90080-3_2

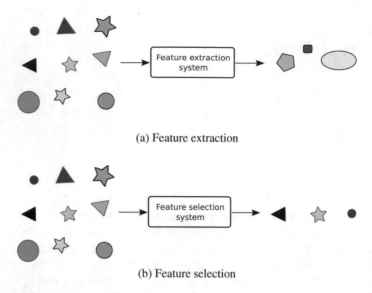

(a) Feature extraction

(b) Feature selection

Fig. 2.1   Examples of feature extraction and feature selection systems

is called Principal Component Analysis (PCA) [1]. This kind of methods are usually applied on fields such as image analysis, signal processing or information retrieval, in which model interpretation is not really important. In turn, feature selection works by keeping the original features that are relevant to the learning task and removing those irrelevant or redundant (see Fig. 2.1b). Since this approach maintains the original features, it is particularly useful for those applications in which model interpretability and knowledge extraction are important, such as, for instance, text mining. The rest of this chapter will be focused on feature selection.

2.1   Foundations of Feature Selection

As mentioned above, feature selection can be defined as the process of detecting the relevant features and discarding the irrelevant ones. A correct selection of the features can lead to an improvement of the inductive learner, either in terms of learning speed, generalization capacity or simplicity of the induced model. Moreover, there are some other benefits associated with a smaller number of features: a reduced measurement cost and hopefully a better understanding of the domain.

There are several situations that can hinder the process of feature selection, such as the presence of irrelevant and redundant features, noise in the data or interaction between attributes. In the presence of hundreds or thousands of features, such as DNA microarray analysis, some researchers notice [2, 3] that is common that a large number of features is not informative because they are either irrelevant or redundant with respect to the class concept. Moreover, when the number of features is high but the number of samples is small, machine learning gets particularly difficult, since the search space will be sparsely populated and the model will not be able to distinguish correctly the relevant data and the noise [4].

There are two typical ways of categorizing feature selection methods. The first one depends on the outcome of the feature selector: whether it returns a subset of relevant features or an ordered ranking of *all* the features, according to their relevance. The first approach is known as *subset evaluation* and the latter as *individual evaluation* or *feature ranking*. In Fig. 2.2, we can see an example of these two approaches. In the case of subset evaluation (Fig. 2.2a) only a subset of the original features is returned by the system —in this example, four out of nine possible features. In turn, a feature ranking system (Fig. 2.2b) returns all the features ranked starting from the strongly relevant features and ending with the weakly relevant ones. In this latter case, it is necessary to establish a threshold in order to reduce the dimensionality of the problem. Most studies in the literature use thresholds that retain different fixed percentages of features [30, 34]. Since threshold values are dependent on the particular dataset being studied, several attempts have been made to develop a general automatic threshold [18, 33, 35]. This issue will be explored in more detail in Chap. 4.

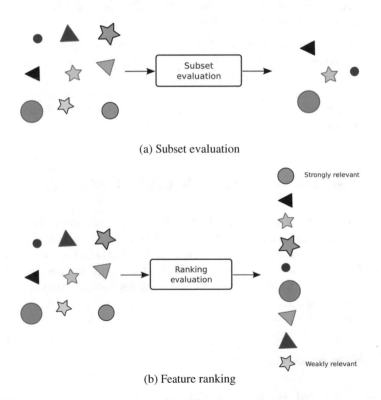

(a) Subset evaluation

(b) Feature ranking

Fig. 2.2 Examples of subset evaluation and feature ranking approaches

Apart from this classification, feature selection methods can also be divided regarding the relationship between a feature selection algorithm and the inductive learning method used to infer a model, into three major approaches [5]:

- *Filters*, which rely on the general characteristics of training data and carry out the feature selection process as a pre-processing step with independence of the induction algorithm. This model is advantageous for its low computational cost and good generalization ability.
- *Wrappers*, which involve a learning algorithm as a black box and consists of using its prediction performance to assess the relative usefulness of subsets of variables. In other words, the feature selection algorithm uses the learning method as a subroutine with the computational burden that comes from calling the learning algorithm to evaluate each subset of features. However, this interaction with the classifier tends to give better performance results than filters.
- *Embedded methods*, which perform feature selection in the process of training and are usually specific to given learning machines. Therefore, the search for an optimal subset of features is built into the classifier construction and can be seen as a search in the combined space of feature subsets and hypotheses. In other words, ensemble methods learn which features best contribute to the accuracy of the model while the model is being created. This approach is able to capture dependencies at a lower computational cost than wrappers.

As can be seen above, each model has its advantages and disadvantages. If the data is not too big, wrappers and embedded methods tend to give more accurate results, at the expense of a high computational cost. Therefore, if speed is crucial (for example in real-time applications) or if we are dealing with Big Data, filters are usually preferred. The interested reader can find more general information about feature selection in the specialized literature [5–8].

2.2 State-of-the-Art Feature Selection Methods

Each year, new feature selection methods are constantly appearing. However, this abundance of feature selection algorithms has not facilitated the choice of a particular method in a given situation, but quite the opposite. Nevertheless, despite the big amount of feature selection methods available, some of them have been able to stand out and their use has become very popular among researchers. Some of them are subsequently described.

2.2.1 Filter Methods

- **Correlation-based Feature Selection** (CFS) is a simple multivariate filter algorithm that ranks feature subsets according to a correlation based heuristic evaluation function [9]. The bias of the evaluation function is toward subsets that contain features that are highly correlated with the class and uncorrelated with each other. Irrelevant features should be ignored because they will have low correlation with the class. Redundant features should be screened out as they will be highly correlated with one or more of the remaining features. The acceptance of a feature will depend on the extent to which it predicts classes in areas of the instance space not already predicted by other features

- The **Consistency-based Filter** [10] evaluates the worth of a subset of features by the level of consistency in the class values when the training instances are projected onto the subset of attributes.
- The **INTERACT** algorithm [11] is a subset filter based on symmetrical uncertainty (SU) and the consistency contribution, which is an indicator about how significantly the elimination of a feature will affect consistency. The algorithm consists of two major parts. In the first part, the features are ranked in descending order based on their SU values. In the second part, features are evaluated one by one starting from the end of the ranked feature list. If the consistency contribution of a feature is less than an established threshold, the feature is removed, otherwise it is selected.
- **Information Gain** [12] is one of the most common attribute evaluation methods. This univariate filter provides an ordered ranking of all the features and then a threshold is required.
- **ReliefF** [13] is an extension of the original Relief algorithm [14]. The original Relief works by randomly sampling an instance from the data and then locating its nearest neighbor from the same and opposite class. The values of the attributes of the nearest neighbors are compared to the sampled instance and used to update relevance scores for each attribute. The rationale is that an useful attribute should differentiate between instances from different classes and have the same value for instances from the same class. ReliefF adds the ability of dealing with multiclass problems and is also more robust and capable of dealing with incomplete and noisy data. This method may be applied in all situations, has low bias, includes interaction among features and may capture local dependencies which other methods miss.
- The **mRMR** (minimum Redundancy Maximum Relevance) method [15] selects features that have the highest relevance with the target class and are also minimally redundant, i.e., selects features that are maximally dissimilar to each other. Both optimization criteria (Maximum-Relevance and Minimum-Redundancy) are based on mutual information.
- **ChiSquared** is a univariate filter based on the χ^2 statistic [36] which evaluates each feature independently with respect to the classes. The higher chi-squared, the more relevant is the feature with respect to class.
- The **FCBF** (Fast Correlation-Based Filter) [3] is a multivariate algorithm that measures feature-class and feature-feature correlation. FCBF starts by selecting a set of features that is highly correlated with the class based on SU. Then, it applies three heuristics that remove the redundant features and keep the feature

that is more relevant to the class. FCBF was designed for high-dimensionality data and has been shown to be effective in removing both irrelevant and redundant features. However, it fails to take into consideration the interaction between features.

2.2.2 Embedded Methods

- **SVM-RFE** (Recursive Feature Elimination for Support Vector Machines) was introduced by Guyon in [16]. This embedded method performs feature selection by iteratively training a SVM classifier with the current set of features and removing the least important feature indicated by the SVM.
- **FS-P** (Feature Selection - Perceptron) [18] is an embedded method based on a perceptron. A perceptron is a type of artificial neural network that can be seen as the simplest kind of feedforward neural network: a linear classifier. The basic idea of this method consists on training a perceptron in the context of supervised learning. The interconnection weights are used as indicators of which features could be the most relevant and provide a ranking.

2.2.3 Wrapper Methods

- **WrapperSubsetEval** [19] evaluates attribute sets by using a learning scheme. Cross validation is used to estimate the accuracy of the learning scheme for a set of attributes. The algorithm starts with the empty set of attributes and searches forward, adding attributes until performance does not improve further. It can be used in conjunction with any learning algorithm.

2.3 Which Is the Best Feature Selection Method?

The benefits of feature selection as a preprocessing step are more than proved in the specialized literature. However, researchers agree that there is not a so-called "best feature selection method" and their efforts are focused on finding a good method for a specific problem setting.

Given the large amount of feature selection methods available, carrying out a comparative study is an arduous task. Moreover, an important issue is how to evaluate the performance of the feature selection methods. If we use the classification accuracy, this is dependent on the particular classifier chosen, and can vary notably from one method to another. Therefore, a possible solution is to evaluate feature selection methods on artificially generated data, in which the desired output is already known. In this way, a feature selection algorithm can be evaluated with independence of the classifier chosen. Citing Belanche et al. [20], there are two main reasons to choose artificial data as a first step to compare feature selection methods:

1. Controlled experiments can be developed by systematically varying chosen experimental conditions, like adding more irrelevant features or noise in the input. This fact facilitates to draw more useful conclusions and to test the strengths and weaknesses of the existing algorithms.
2. The main advantage of artificial scenarios is the knowledge of the set of optimal features that must be selected, thus the degree of closeness to any of these solutions can be assessed in a confident way.

In the rest of this section, we will compare the state-of-the-art feature selection methods described in Sect. 2.2 using a set of 11 synthetic datasets covering a large suite of problems (non-linearity of the data, noise in the inputs and in the target, increasing number of irrelevant and redundant features, etc.).

2.3.1   Datasets

As mentioned above, we will test and compare state-of-the-art feature selection methods on 11 synthetic datasets:

- CorrAL [21] is a classical dataset with six binary features (f_1, f_2, f_3, f_4, f_5, f_6) and the class value is $(f_1 \wedge f_2) \vee (f_3 \wedge f_4)$. Feature f_5 is irrelevant and f_6 is correlated to the class label by 75%. We also included CorrAL-100 [32] which was constructed by adding 93 irrelevant binary features.
- XOR-100 was constructed from the original XOR problem. In this case, the dataset has 2 relevant features that are both necessary to define the class (class equals $f_1 \oplus f_2$) and the remaining 97 binary features are irrelevant.
- Parity3+3 is a classic problem where the output is $f(x_1, \ldots, x_n) = 1$ if the number of $x_i = 1$ is odd and $f(x_1, \ldots, x_n) = 0$ otherwise. In this case, the target concept is the parity of three bits. It contains 12 features among which 3 are relevant, another 3 are redundant (repeated) and other 6 are irrelevant (randomly generated).
- The LED problem [24] is a simple classification task that consists of, given the active leds on a seven segments display, identifying the digit that the display is representing. Thus, the classification task to be solved is described by seven binary attributes and ten possible classes available ($C = \{0, 1, 2, 3, 4, 5, 6, 7, 8, 9\}$). A 1 in an attribute indicates that the led is active, and a 0 indicates that it is not active. Two versions of the Led problem will be used: the first one, Led25, adding 17 irrelevant attributes (with random binary values) and the second one, Led100, adding 92 irrelevant attributes. Both versions contain 50 samples. The small number of

samples was chosen because we are interested in dealing with datasets with a high number of features and a small sample size. Besides, different levels of noise (altered inputs) have been added to the attributes of these two versions of the Led dataset: 2, 6, 10, 15 and 20%. In this manner, the tolerance to different levels of noise of the feature selection methods tested will be checked. Note that, as the attributes take binary values, adding noise means assigning to the relevant features an incorrect value.

- Monk3 belongs to the collection of MONK's problems [23] in which robots are described by six different attributes $(x_1, \ldots, x_6)$. It is a binary classification task in which the class is $(x_5 = 3 \wedge x_4 = 1) \vee (x_5 \neq 4 \wedge x_2 \neq 3)$. Among the 122 samples, 5% are misclassifications, i.e. noise in the target.
- The SD datasets [22] represent the challenging problematic of microarray data, with a high number of features whilst small number of samples, besides of a high number of irrelevant and redundant features. SD1, SD2 and SD3 are three-class datasets with 75 samples (each class containing 25 samples) and 4000 irrelevant features. SD1 contains two groups of relevant genes generated from a multivariate normal distribution, with 10 genes in each group. Genes in the same group are redundant with each other and the optimal gene subset for distinguishing the three classes consists of any two relevant genes from different groups. In turn, SD2 contains four groups of 10 relevant genes; and SD3 contains six groups of 10 relevant genes.
- The Madelon dataset [5] is a 2 class problem originally proposed in the NIPS'2003 feature selection challenge. The relevant features are situated on the vertices of a five dimensional hypercube. Five redundant features were added, obtained by multiplying the useful features by a random matrix. Some of the previously defined features were repeated to create 10 more features. The other 480 features are drawn from a Gaussian distribution and labeled randomly. This dataset presents high dimensionality both in number of features and in number of samples and the data were distorted by adding noise, flipping labels, shifting and rescaling.

2.3.2 Experimental Study

Some of the state-of-the-art feature selection methods described in Sect. 2.2 are tested and compared trying to draw useful conclusions. As mentioned above, feature selection methods can return a subset of features or a rankings of all features, in which case it is necessary to establish a threshold. In these experiments, we heuristically set the following rules to decide the number of features that ranker methods should return, according to the number of total features (N):

- if N < 10, select 75% of features
- if 10 < N < 75, select 40% of features
- if 75 < N < 100, select 10% of features
- if N > 100, select 3% of features

According to these rules, the number of features that will be returned by ranker methods is 5 for the datasets Corral, Parity3+3 and Monk3, 10 for the datasets Corral-100, XOR-100 and Led, and 15 for Madelon.

A scoring measure was defined in order to fairly compare the effectiveness showed by the different feature selection methods. This measure is an index of success *suc.*, see (2.1), which attempts to reward the selection of relevant features and to penalize the inclusion of irrelevant ones, penalizing two situations:

- The solution is *incomplete*: there are relevant features lacking.
- The solution is *incorrect*: there are some irrelevant features.

$$ suc. = \left[\frac{R_s}{R_t} - \alpha \frac{I_s}{I_t} \right] \times 100, \tag{2.1} $$

where R_s is the number of relevant features selected, R_t is the total number of relevant features, I_s is the number of irrelevant features selected and I_t is the total number of irrelevant features. The term α was introduced to ponder that choosing an irrelevant feature is better than missing a relevant one (i.e. we prefer an incorrect solution rather than an incomplete one). The parameter α is defined as $\alpha = \min\{\frac{1}{2}, \frac{R_t}{I_t}\}$. Note that the higher the success, the better the method, and 100 is the maximum.

In the case of ranker methods and in order to be fair, if all the optimal features are selected before any irrelevant feature, the index of success will be 100, due to the fact that the number of features that ranker methods are forced to select is always larger than the number of relevant features.

As was explained above, the evaluation of the feature selection methods is done by counting the number of correct/wrong features. However, it is also interesting and a common practice in the literature [25] to see the classification accuracy obtained in a 10-fold cross-validation, in order to check if the true model (that is, the one with an index of success of 100) is also unique (that is, if is the only one that can achieve the best percentage of classification success). For this purpose, four well-known classifiers, based on different models, were chosen: C4.5 [26], naive Bayes (NB) [27], IB1 [28] and SVM [29]. Notice that the embedded feature selection method SVM-RFE is tested both with a linear kernel and with a radial basis function (RBF) kernel.

In Tables 2.1 and 2.2 we study how the feature selection methods can deal with correlation and redundancy, using the datasets Corral and Corral-100. In both cases, the desired behavior of a feature selection method is to select the 4 relevant features and to discard the irrelevant ones, as well as detecting the correlated feature and not selecting it. Regarding Corral-100, it is curious that the best classification accuracy was obtained by SVM-RFE-linear, which has a index of success of 25, but this fact can be explained because in this dataset there are some irrelevant features that are informative to the classifiers (see more details in [30]).

Table 2.1 Results for CorrAL. "Rel." shows the relevant features selected, "C" indicates if the correlated feature is selected (✓) or not (✗), "Irr." means the number of irrelevant features selected and "suc." represents the index of success

Method	Rel.	C	Irr.	suc.	Accuracy (%)			
					C4.5	NB	IB1	SVM
CFS	–	✓	0	-25	75.00	75.00	59.38	75.00
Consistency	–	✓	0	-25	75.00	75.00	59.38	75.00
INTERACT	–	✓	0	-25	75.00	75.00	59.38	75.00
InfoGain	–	✓	0	-25	75.00	75.00	59.38	75.00
ReliefF	1-4	✓	0	75	62.50	81.25	96.88	87.50
mRMR	1-4	✓	0	75	62.50	81.25	96.88	87.50
SVM-RFE-linear	1-4	✓	0	75	62.50	81.25	96.88	87.50
SVM-RFE-RBF	1-4	✗	1	75	81.25	78.13	81.25	71.86
FS-P	1-4	✗	0	**100**	81.25	78.13	**100.00**	81.25
Wrapper SVM	–	✓	0	-25	75.00	75.00	59.38	75.00
Wrapper C4.5	–	✓	0	-25	75.00	75.00	59.38	75.00

Table 2.2 Results for CorrAL-100. "Rel." shows the relevant features selected, "C" indicates if the correlated feature is selected (✓) or not (✗), "Irr." means the number of irrelevant features selected and "suc." represents the index of success

Method	Rel.	C	Irr.	suc.	Accuracy (%)			
					C4.5	NB	IB1	SVM
CFS	–	✓	0	-2	75.00	75.00	59.38	75.00
Consistency	–	✓	0	-2	75.00	75.00	59.38	75.00
INTERACT	–	✓	0	-2	75.00	75.00	59.38	75.00
InfoGain	–	✓	0	-2	75.00	75.00	59.38	75.00
ReliefF	1-3	✓	6	75	53.13	84.38	87.50	81.25
mRMR	1-4	✓	5	**99**	53.13	81.25	90.63	90.63
SVM-RFE-linear	4	✓	8	25	62.50	87.50	68.75	**96.88**
SVM-RFE-RBF	–	✓	9	-44	68.75	68.75	62.50	75.00
FS-P	1,3,4	✓	6	75	53.13	87.50	84.38	87.50
Wrapper SVM	–	✓	0	-2	75.00	75.00	59.38	75.00
Wrapper C4.5	–	✓	2	-13	84.38	75.00	75.00	75.00

To analyze the behavior of feature selection methods when dealing with non-linearity, Tables 2.3 and 2.4 present the results on the datasets XOR and Parity. XOR-100 contains 2 relevant features and 97 irrelevant features whilst Parity has 3 relevant, 3 redundant and 6 irrelevant features. For the sake of completeness, SVM and naive Bayes will be applied over these two datasets. However, bearing in mind that those methods are linear classifiers (a linear kernel is being used for SVM), no good results are to be expected.

Table 2.3 Results for XOR-100. "Rel." shows the relevant features selected, "Red" indicates the number of redundant features elected, "Irr." means the number of irrelevant features selected and "suc." represents the index of success

Method	Rel.	Irr.	suc.	Accuracy (%)			
				C4.5	NB	IB1	SVM
ReliefF	1,2	0	100	100.00	64.00	100.00	70.00
mRMR	1	9	50	52.00	74.00	64.00	72.00
SVM-RFE-linear	–	10	-21	48.00	68.00	56.00	78.00
SVM-RFE-RBF	1,2	0	100	100.00	64.00	100.00	70.00
FS-P	1	9	50	62.00	76.00	62.00	74.00
Wrapper SVM	–	1	-2	66.00	66.00	60.00	66.00
Wrapper C4.5	1,2	2	99	100.00	70.00	96.00	50.00

Table 2.4 Results for Parity3+3. "Rel." shows the relevant features selected, "Red" indicates the number of redundant features elected, "Irr." means the number of irrelevant features selected and "suc." represents the index of success

Method	Rel.	Red.	Irr.	suc.	Accuracy (%)			
					C4.5	NB	IB1	SVM
ReliefF	1,2,3	2	0	93	90.63	29.69	100.00	37.50
mRMR	2,3	0	3	56	60.94	59.38	59.38	59.38
SVM-RFE-linear	3	0	4	19	54.69	59.38	46.88	57.81
SVM-RFE-RBF	1,2,3	0	0	100	90.63	31.25	100.00	25.00
FS-P	–	0	5	-19	51.56	57.81	56.25	57.81
Wrapper SVM	–	0	1	-4	64.06	64.06	57.81	64.06
Wrapper C4.5	–	0	1	-4	64.06	64.06	57.81	64.06

As can be seen, the methods CFS, Consistency, INTERACT and InfoGain do not appear because they were not able to solve these non-linear problems, so they returned an empty subset of features. On the other hand, the filter ReliefF and the embedded method SVM-RFE-RBF detected the relevant features both in XOR-100 and in Parity3+3, achieving the best indices of success and leading to high classification accuracy.

Tables 2.5 and 2.6 show how the different feature selection methods are tolerant to noise in the input, using Led-25 and Led-100 datasets with different levels of noise (2, 6, 10, 15 and 20%). It has to be noted that, as the attributes take binary values, adding noise means assigning to the relevant features an incorrect value. As expected, in general the index of success decreases when the level of noise increases, and worse performances were obtained over Led-100 due to the higher number of irrelevant features. Regarding the wrapper model, both versions tested degrade their results with the presence of noise, both in Led-25 and Led-100. The embedded method FS-P shows a strange behavior on Led-25, since it degrades slightly in terms of index

Table 2.5 Results for Led-25 dataset with different levels of noise (N) in inputs

N(%)	Method	Relevant	Irr. No.	suc.	Accuracy (%)			
					C4.5	NB	IB1	SVM
0	CFS	1-5,7	0	86	92.00	**100.00**	**100.00**	96.00
	Consistency	1-5	0	71	92.00	**100.00**	**100.00**	96.00
	INTERACT	1-5,7	0	86	92.00	**100.00**	**100.00**	96.00
	InfoGain	1-7	0	**100**	92.00	**100.00**	**100.00**	96.00
	ReliefF	1-7	3	99	92.00	94.00	96.00	**100.00**
	mRMR	1-5,7	4	85	92.00	94.00	88.00	96.00
	SVM-RFE-linear	3-7	5	71	46.00	54.00	48.00	48.00
	SVM-RFE-RBF	1-6	4	85	92.00	92.00	80.00	94.00
	FS-P	1-7	3	99	92.00	92.00	86.00	96.00
	Wrapper SVM	1-5	2	71	92.00	90.00	82.00	**100.00**
	Wrapper C4.5	1-5	0	71	92.00	**100.00**	**100.00**	96.00
2	CFS	1-5	0	71	90.00	**98.00**	96.00	94.00
	Consistency	1-5	0	71	90.00	**98.00**	96.00	94.00
	INTERACT	1-5	0	71	90.00	**98.00**	96.00	94.00
	InfoGain	1-7	0	**100**	90.00	96.00	94.00	94.00
	ReliefF	1-7	3	99	90.00	90.00	84.00	92.00
	mRMR	1-5,7	4	85	88.00	86.00	80.00	86.00
	SVM-RFE-linear	3-7	5	71	68.00	70.00	54.00	70.00
	SVM-RFE-RBF	1-6	4	85	90.00	90.00	74.00	88.00
	FS-P	1-7	3	99	90.00	86.00	82.00	90.00
	Wrapper SVM	1-5	2	71	90.00	88.00	80.00	96.00
	Wrapper C4.5	1-5	0	71	90.00	**98.00**	96.00	94.00
6	CFS	1,2,4,5,7	0	71	72.00	78.00	72.00	70.00
	Consistency	1,2,4,5,7	0	71	72.00	78.00	72.00	70.00
	INTERACT	1,2,4,5,7	0	71	72.00	78.00	72.00	70.00
	InfoGain	1,2,4,5,7	0	71	72.00	78.00	72.00	70.00
	ReliefF	1-5,7	4	85	60.00	66.00	68.00	72.00
	mRMR	1-5,7	4	85	60.00	66.00	68.00	72.00
	SVM-RFE-linear	2,3,5	7	42	52.00	50.00	34.00	52.00
	SVM-RFE-RBF	1-6	4	85	70.00	72.00	50.00	72.00
	FS-P	1-6	4	85	72.00	56.00	62.00	70.00
	Wrapper SVM	1-7	15	99	56.00	54.00	58.00	**84.00**
	Wrapper C4.5	1,2,4,5	2	57	76.00	72.00	66.00	72.00
10	CFS	1,2,4,7	0	57	60.00	50.00	58.00	46.00
	Consistency	1,2,4,7	0	57	60.00	50.00	58.00	46.00
	INTERACT	1,2,4,7	0	57	60.00	50.00	58.00	46.00
	InfoGain	1,2,4,7	0	57	60.00	50.00	58.00	46.00
	ReliefF	1,2,4,5,7	5	71	74.00	54.00	66.00	64.00
	mRMR	1,2,4,5,7	5	71	66.00	60.00	66.00	66.00
	SVM-RFE-linear	2,3,5,7	6	57	44.00	36.00	38.00	42.00
	SVM-RFE-RBF	1,3,5	7	42	26.00	34.00	30.00	40.00
	FS-P	1-6	4	**85**	60.00	46.00	48.00	58.00
	Wrapper SVM	1,2,4	9	42	72.00	56.00	56.00	**78.00**
	Wrapper C4.5	1,2,4	3	43	76.00	58.00	56.00	66.00

Table 2.5   (continued)

	Method							
	CFS	1,7	0	29	28.00	28.00	32.00	36.00
	Consistency	1,7	0	29	28.00	28.00	32.00	36.00
	INTERACT	1,7	0	29	28.00	28.00	32.00	36.00
	InfoGain	1,7	0	29	28.00	28.00	32.00	36.00
	ReliefF	1,2,4,5,7	5	71	54.00	50.00	54.00	**64.00**
15	mRMR	1,2,4,5,7	5	71	54.00	50.00	54.00	**64.00**
	SVM-RFE-linear	3,5,7	7	42	30.00	20.00	16.00	26.00
	SVM-RFE-RBF	1,5	8	28	16.00	24.00	12.00	16.00
	FS-P	1,3,5,6,7	5	71	30.00	28.00	22.00	26.00
	Wrapper SVM	1,2,6	5	42	50.00	50.00	42.00	**64.00**
	Wrapper C4.5	1,2,5,7	2	57	58.00	50.00	46.00	52.00
	CFS	1	0	14	28.00	20.00	28.00	28.00
	Consistency	1	0	14	28.00	20.00	28.00	28.00
	INTERACT	1	0	14	28.00	20.00	28.00	28.00
	InfoGain	1	0	14	28.00	20.00	28.00	28.00
	ReliefF	1,2,5,7	6	57	30.00	38.00	44.00	44.00
20	mRMR	1,2,5,7	6	57	34.00	38.00	42.00	**48.00**
	SVM-RFE-linear	–	10	-1	8.00	26.00	20.00	20.00
	SVM-RFE-RBF	1,2,3,5	6	57	32.00	32.00	14.00	26.00
	FS-P	1-3,5,6	5	71	18.00	24.00	24.00	20.00
	Wrapper SVM	1	3	14	36.00	38.00	28.00	44.00
	Wrapper C4.5	1,5	4	28	44.00	32.00	28.00	32.00

of success (from 100 to 93%), but the degradation in classification accuracy is more than notable (from 92–100% to 34–40%), and a similar situation happens with Led-100. This can be explained by the fact that adding noise changes the information contained by the relevant features that can be not relevant anymore, and not helping the classification process. A similar situation happens with the filters mRMR and ReliefF, which are robust to noise in terms of index of success, but the classification accuracy is more affected (although they are usually obtained the highest accuracies with high levels of noise). To sum up, the filters mRMR and ReliefF and the embedded method FS-P are the methods most tolerant to noise in the inputs and the subsets filters (CFS, Consistency and INTERACT) and Information Gain are the most affected by noise.

Table 2.7 shows the results for the Monk3 problem, which includes a 5% of mis-classifications, i.e. noise in the target. The relevant features are x_2, x_4 and x_5. However, as it was stated in [31], for a feature selection algorithm it is easy to find the variables x_2 and x_5, which together yield the second conjunction in the expression seen in Sect. 2.3.1. According to the experimental results presented in [31], selecting those features can lead to a better performance than selecting the three relevant ones. This additional information can help to explain the fact that in Table 2.7 several algorithms selected only two of the relevant features.

Table 2.6 Results for Led-100 dataset with different levels of noise (N) in inputs

N(%)	Method	Relevant	Irr. No.	suc.	Accuracy (%)			
					C4.5	NB	IB1	SVM
0	CFS	1-5,7	0	86	92.00	**100.00**	**100.00**	96.00
	Consistency	1-5	0	71	92.00	**100.00**	**100.00**	96.00
	INTERACT	1-5,7	0	86	92.00	**100.00**	**100.00**	96.00
	InfoGain	1-7	0	**100**	92.00	**100.00**	**100.00**	96.00
	ReliefF	1-7	3	99	92.00	94.00	96.00	**100.00**
	mRMR	1-5,7	4	85	92.00	94.00	88.00	96.00
	SVM-RFE-linear	3-7	5	71	46.00	54.00	48.00	48.00
	SVM-RFE-RBF	1-6	4	85	92.00	92.00	80.00	94.00
	FS-P	1-7	3	99	92.00	92.00	86.00	96.00
	Wrapper SVM	1-5	2	71	92.00	90.00	82.00	**100.00**
	Wrapper C4.5	1-5	0	71	92.00	**100.00**	**100.00**	96.00
2	CFS	1-5	0	71	90.00	**98.00**	96.00	94.00
	Consistency	1-5	0	71	90.00	**98.00**	96.00	94.00
	INTERACT	1-5	0	71	90.00	**98.00**	96.00	94.00
	InfoGain	1-7	0	**100**	90.00	96.00	94.00	94.00
	ReliefF	1-7	3	99	90.00	90.00	84.00	92.00
	mRMR	1-5,7	4	85	88.00	86.00	80.00	86.00
	SVM-RFE-linear	3-7	5	71	68.00	70.00	54.00	70.00
	SVM-RFE-RBF	1-6	4	85	90.00	90.00	74.00	88.00
	FS-P	1-7	3	99	90.00	86.00	82.00	90.00
	Wrapper SVM	1-5	2	71	90.00	88.00	80.00	96.00
	Wrapper C4.5	1-5	0	71	90.00	**98.00**	96.00	94.00
6	CFS	1,2,4,5,7	0	71	72.00	78.00	72.00	70.00
	Consistency	1,2,4,5,7	0	71	72.00	78.00	72.00	70.00
	INTERACT	1,2,4,5,7	0	71	72.00	78.00	72.00	70.00
	InfoGain	1,2,4,5,7	0	71	72.00	78.00	72.00	70.00
	ReliefF	1-5,7	4	85	60.00	66.00	68.00	72.00
	mRMR	1-5,7	4	85	60.00	66.00	68.00	72.00
	SVM-RFE-linear	2,3,5	7	42	52.00	50.00	34.00	52.00
	SVM-RFE-RBF	1-6	4	85	70.00	72.00	50.00	72.00
	FS-P	1-6	4	85	72.00	56.00	62.00	70.00
	Wrapper SVM	1-7	15	**99**	56.00	54.00	58.00	**84.00**
	Wrapper C4.5	1,2,4,5	2	57	76.00	72.00	66.00	72.00
10	CFS	1,2,4,7	0	57	60.00	50.00	58.00	46.00
	Consistency	1,2,4,7	0	57	60.00	50.00	58.00	46.00
	INTERACT	1,2,4,7	0	57	60.00	50.00	58.00	46.00
	InfoGain	1,2,4,7	0	57	60.00	50.00	58.00	46.00
	ReliefF	1,2,4,5,7	5	71	74.00	54.00	66.00	64.00
	mRMR	1,2,4,5,7	5	71	66.00	60.00	66.00	66.00
	SVM-RFE-linear	2,3,5,7	6	57	44.00	36.00	38.00	42.00
	SVM-RFE-RBF	1,3,5	7	42	26.00	34.00	30.00	40.00
	FS-P	1-6	4	**85**	60.00	46.00	48.00	58.00
	Wrapper SVM	1,2,4	9	42	72.00	56.00	56.00	**78.00**
	Wrapper C4.5	1,2,4	3	43	76.00	58.00	56.00	66.00

Table 2.6 (continued)

	Method							
	CFS	1,7	0	29	28.00	28.00	32.00	36.00
	Consistency	1,7	0	29	28.00	28.00	32.00	36.00
	INTERACT	1,7	0	29	28.00	28.00	32.00	36.00
	InfoGain	1,7	0	29	28.00	28.00	32.00	36.00
	ReliefF	1,2,4,5,7	5	71	54.00	50.00	54.00	**64.00**
15	mRMR	1,2,4,5,7	5	71	54.00	50.00	54.00	**64.00**
	SVM-RFE-linear	3,5,7	7	42	30.00	20.00	16.00	26.00
	SVM-RFE-RBF	1,5	8	28	16.00	24.00	12.00	16.00
	FS-P	1,3,5,6,7	5	71	30.00	28.00	22.00	26.00
	Wrapper SVM	1,2,6	5	42	50.00	50.00	42.00	**64.00**
	Wrapper C4.5	1,2,5,7	2	57	58.00	50.00	46.00	52.00
	CFS	1	0	14	28.00	20.00	28.00	28.00
	Consistency	1	0	14	28.00	20.00	28.00	28.00
	INTERACT	1	0	14	28.00	20.00	28.00	28.00
	InfoGain	1	0	14	28.00	20.00	28.00	28.00
	ReliefF	1,2,5,7	6	57	30.00	38.00	44.00	44.00
20	mRMR	1,2,5,7	6	57	34.00	38.00	42.00	**48.00**
	SVM-RFE-linear	–	10	-1	8.00	26.00	20.00	20.00
	SVM-RFE-RBF	1,2,3,5	6	57	32.00	32.00	14.00	26.00
	FS-P	1-3,5,6	5	71	18.00	24.00	24.00	20.00
	Wrapper SVM	1	3	14	36.00	38.00	28.00	44.00
	Wrapper C4.5	1,5	4	28	44.00	32.00	28.00	32.00

Studying the index of success in Table 2.7, one can see that only ReliefF achieved a value of 100. The worst behavior was shown by mRMR, since it selected the three irrelevant features. As was justified above, many methods selected only two of the relevant features and it can be considered a good comportment. For IB1 classifier, the best accuracy corresponds to ReliefF, which also obtained the best result in terms of index of success.

Another problematic we want to test is the case in which the number of features is much larger than the number of samples. This situation is represented by datasets SD1, SD2 and SD3. For these datasets, besides of using the index of success and classification accuracy, we will use the measures employed in [22], which are more specific for this problem. Hence, the performance on datasets SD1, SD2 and SD3 (see Tables 2.8, 2.9 and 2.10) will be also evaluated in terms of:

- (**#**): number of selected features.
- **OPT(x)**: number of selected features within the optimal subset, where x indicates the optimal number of features.
- **Red**: number of redundant features.
- **Irr**: number of irrelevant features.

Table 2.7 Results for Monk3. Relevant features: 2,4,5

Method	Relevant	Irr. No.	suc.	Accuracy (%)			
				C4.5	NB	IB1	SVM
CFS	2,5	0	67	**93.44**	88.52	89.34	79.51
Consistency	2,5	0	67	**93.44**	88.52	89.34	79.51
INTERACT	2,5	0	67	**93.44**	88.52	89.34	79.51
InfoGain	2,5	0	67	**93.44**	88.52	89.34	79.51
ReliefF	2,5,4	0	**100**	**93.44**	88.52	90.98	80.33
mRMR	2,5	3	17	92.62	88.52	80.33	78.69
SVM-RFE-linear	2,4,5	2	67	**93.44**	88.52	84.43	84.43
SVM-RFE-RBF	2,4,5	2	67	**93.44**	88.52	84.43	84.43
FS-P	2,4,5	2	67	**93.44**	88.52	84.43	84.43
Wrapper SVM	2,4,5	1	83	**93.44**	89.34	82.79	79.51
Wrapper C4.5	2,5	0	67	**93.44**	88.52	89.34	79.51

Table 2.8 Features selected by each algorithm on synthetic dataset SD1. Ranker methods are tested selecting the optimal number and 20 features as cardinality

	(#)	OPT(2)	Red	Irr	suc	Accuracy (%)			
						C4.5	NB	IB1	SVM
CFS	28	2	1	25	**100**	57.33	82.67	69.33	77.33
Cons	8	2	0	6	**100**	54.67	76.00	60.00	66.67
INT	23	2	0	21	**100**	60.00	81.33	66.67	80.00
IG	42	2	15	25	**100**	58.67	72.00	70.67	78.67
ReliefF[1]	2	1	1	0	50	40.00	45.33	44.00	46.67
ReliefF[2]	20	2	13	5	**100**	60.00	61.33	70.67	73.33
mRMR[1]	2	1	0	1	50	41.33	49.33	34.67	50.67
mRMR[2]	20	1	0	19	50	54.67	82.67	68.00	78.67
SVM-RFE-linear[1]	2	2	0	0	**100**	56.00	60.00	52.00	57.33
SVM-RFE-linear[2]	20	2	3	15	**100**	46.67	88.00	76.00	92.00
FS-P[1]	2	0	0	2	0	37.33	49.33	41.33	50.67
FS-P[2]	20	1	2	17	50	53.33	76.00	65.33	73.33
W-SVM	19	1	0	18	50	44.00	74.67	58.67	**94.67**
W-C4.5	10	0	0	10	0	77.33	38.67	40.00	38.67

[1] Selecting the optimal number of features.

[2] Selecting 20 features.

For the ranker methods ReliefF, mRMR, SVM-RFE and FS-P, two different cardinalities were tested: the optimal number of features and 20, since the subset methods have a similar cardinality. It has to be noted that in this problem and for the calculation of the index of success, redundant features are treated the same as irrelevant features in Eq. (2.1). Notice that the index of success is 100 even with 25 irrelevant features selected, due to the high number of irrelevant features (4000).

Studying the selected features, the subset filters and InfoGain (which exhibits a similar behavior) showed excellent results, in all SD1, SD2 and SD3. Also SVM-RFE obtained good results, although the version with the RBF kernel could not been

Table 2.9 Features selected by each algorithm on synthetic dataset SD2. Ranker methods are tested selecting the optimal number and 20 features as cardinality

	(#)	OPT(4)	Red	Irr	suc	Accuracy (%)			
						C4.5	NB	IB1	SVM
CFS	21	4	0	17	100	64.00	**84.00**	72.00	81.33
Cons	9	4	0	5	100	54.67	70.67	60.00	72.00
INT	20	3	0	17	75	70.67	80.00	74.67	81.33
IG	40	4	19	17	100	61.33	69.33	61.33	76.00
ReliefF[1]	4	0	0	4	0	48.00	64.00	50.67	52.00
ReliefF[2]	20	1	9	10	25	54.67	60.00	61.33	70.67
mRMR[1]	4	1	0	3	25	54.67	64.00	60.00	57.33
mRMR[2]	20	1	0	19	25	60.00	70.67	44.00	68.00
SVM-RFE-linear[1]	4	3	1	0	75	46.67	62.67	54.67	65.33
SVM-RFE-linear[2]	20	4	4	12	100	57.33	82.67	69.33	**84.00**
FS-P[1]	4	0	0	20	0	42.67	54.67	40.00	57.33
FS-P[2]	20	0	0	20	0	52.00	68.00	42.67	61.33
W-SVM	13	1	0	12	25	44.00	60.00	45.33	77.33
W-C4.5	6	1	0	5	25	72.00	46.67	34.67	42.67

[1] Selecting the optimal number of features.

[2] Selecting 20 features.

Table 2.10 Features selected by each algorithm on synthetic dataset SD3. Ranker methods are tested selecting the optimal number and 20 features as cardinality

	(#)	OPT(6)	Red	Irr	suc	Accuracy (%)			
						C4.5	NB	IB1	SVM
CFS	23	4	2	17	67	64.00	80.00	73.33	70.67
Cons	9	3	0	6	50	58.67	76.00	62.67	76.00
INT	19	4	1	14	67	61.33	82.67	70.67	66.67
IG	49	4	31	14	67	62.67	65.33	65.33	73.33
ReliefF[1]	6	1	5	0	17	50.67	57.33	45.33	53.33
ReliefF[2]	20	1	9	10	17	56.00	69.33	61.33	68.00
mRMR[1]	6	1	0	5	17	62.67	62.67	66.67	65.33
mRMR[2]	20	1	0	19	17	50.67	77.33	52.00	66.67
SVM-RFE-linear[1]	6	3	0	3	50	56.00	70.67	61.33	65.33
SVM-RFE-linear[2]	20	4	2	14	67	49.33	**85.33**	70.67	82.67
FS-P[1]	6	0	0	6	0	36.00	54.67	34.67	46.67
FS-P[2]	20	1	0	19	17	38.67	61.33	45.33	56.00
W-SVM	10	1	0	9	17	48.00	61.33	61.33	81.33
W-C4.5	5	1	0	4	17	68.00	50.67	37.33	48.00

[1] Selecting the optimal number of features.

[2] Selecting 20 features.

applied on these datasets due to memory complexity. With respect to the classifiers, SVM achieves the highest accuracy.

Finally, we can see the results on Madelon in Table 2.11. This is a very complex artificial dataset which is distorted by adding noise, flipping labels, shifting and rescaling. It is also a non-linear problem, so it conforms a challenge for feature selection researchers. The desired behavior for a feature selection method is to select the relevant features (1–5) and discard the redundant and irrelevant ones. Notice that

Table 2.11 Results for Madelon. Relevant features: 1–5

Method	Relevant	Red. No.	Irr. No.	suc.	Accuracy (%)			
					C4.5	NB	IB1	SVM
CFS	3	7	0	20	80.92	69.58	86.83	66.08
Consistency	3,4	10	0	40	83.54	69.67	**90.83**	66.83
INTERACT	3,4	10	0	40	83.54	69.67	**90.83**	66.83
InfoGain	3,4	10	0	40	83.54	69.67	**90.83**	66.83
ReliefF	1,3,4,5	11	0	80	84.21	69.83	89.88	66.46
mRMR	–	1	14	0	64.92	62.25	53.13	57.08
SVM-RFE-linear	1,3,4,5	4	7	80	86.42	66.88	81.25	67.42
FS-P	3,4	3	10	40	70.50	66.17	62.54	66.96
Wrapper SVM	3	0	16	20	66.63	66.04	54.08	67.54
Wrapper C4.5	1-5	5	15	99	87.04	70.00	75.42	66.33

for the calculation of the index of success, the redundant attributes selected stand for irrelevant features. The results for SVM and naive Bayes will not be analyzed, since they are linear classifiers. The best result in terms of index of success was obtained by the wrapper with C4.5, selecting all the 5 relevant features, which also led to the best classification accuracy for C4.5.

In light of the results presented in this section, the authors suggest some guidelines:

- In complete ignorance of the particulars of data, the authors suggest to use the filter ReliefF. It detects relevance in a satisfactory manner, even in complex datasets such as XOR-100, and it is tolerant to noise (both in the inputs and in the output). Moreover, due to the fact that it is a filter, it has the implicit advantage of its low computational cost.
- When dealing with high non-linearity of data (such as XOR-100 and Parity3+3), SVM-RFE with RBF kernel is an excellent choice, since it is able to solve these complex problems. However, at the expense of being computationally more expensive than the remaining approaches seen in this work.
- In the presence of altered inputs, the best option is to use the embedded method FS-P, since it has proved to be very robust to noise. A less expensive alternative is the use of the filters ReliefF or mRMR, which also shown good behaviors over this scenario. With low levels of noise (up to 6%), the authors also suggest the use of the filter Information Gain.
- When the goal is to select the smallest number of irrelevant features (even at the expense of selecting fewer relevant features), we suggest to employ one of the subset filters (CFS, Consistency-based or INTERACT). This kind of methods have the advantage of releasing the user from the task of deciding how many features to choose.
- When dealing with datasets with a small ratio between number of samples and features and a high number of irrelevant attributes, which is part of the problematics of microarray data, the subset filters and Information Gain presented a promising behavior. SVM-RFE performs also adequately, but because of being an embedded

method is computationally expensive, especially in high-dimensional datasets like these.

- In general, the authors suggest the use of filters (specifically ReliefF), since they carry out the feature selection process with independence of the induction algorithm and are faster than embedded and wrapper methods. However, in case of using another approach, we suggest to use the embedded method FS-P.

2.4 On the Scalability of Feature Selection Methods

Apart from being difficult to decide which is the best feature selection method, as seen in the previous section, another important issue is the scalability of the existing methods. Most algorithms were developed when dataset sizes were much smaller, but nowadays distinct compromises are required for the case of small-scale and large-scale (big data) learning problems. Small-scale learning problems are subject to the usual approximation-estimation trade-off. In the case of large-scale learning problems, the trade-off is more complex because it involves not only the accuracy of the selection but also other aspects. Stability, that is the sensitivity of the results to training set variations, is one of such factors. The other important aspect is scalability, that is the behavior of the algorithms in the case in which the training set is increasingly high. In general, one can say that most of the classical feature selection approaches that are univariate have an important advantage in scalability, but at the cost of ignoring feature dependencies, and thus perhaps leading to lower performances than other feature selection techniques. To improve performance, multivariate techniques are proposed, but at the cost of reducing scalability. In this situation, the scalability of a feature selection method becomes extremely important.

In this section, we will comment on the scalability of some state-of-the-art feature selection methods described in Sect. 2.2, based on a previous work [37]. For analyzing their scalability, new evaluation measures were proposed, which were based not only on the accuracy of the selection, but also on other aspects such as the execution time or the stability of the returned features. The performance of the methods was evaluated, as well as in the previous section, on an artificial controlled experimental scenario.

2.4.1 Experimental Study

The scalability of feature selection methods will be evaluated in terms of (a) error, that is the percentage of selected features that are not relevant for the problem; (b) distance, which is the inverse of the stability, and measures how different two rankings of subsets of features are; and (c) computational time to select the features. Notice that all these measures are desirable to be minimized, and that the error and the distance are bounded between 0 and 1.

Table 2.12 Overview of scalability of filters (notice that the larger the number of dots, the better the behavior)

	Method	Error	Distance	Training time
Ranking	Chi-Squared	•••	••••••	••••••
	ReliefF	••••••	••••	•
	InfoGain	•••	••••••	••••••
	mRMR	•••	•••	••••••
Subset	FCBF	••••	••••	••••
	CFS	••••	••••••	•••
	Consistency	•••	••••	•
	INTERACT	••••	••••	•

In our previous work [37] the interested reader can find more detailed results, but here, for the sake of brevity, only a summary is presented. Table 2.12 shows an overview of the behavior of the different FS methods in the three considered metrics (error, distance and training time), where the larger the number of dots, the better the behavior. The number of dots is calculated according to the following. For each metric, we recall the worst value for all methods and datasets. Each specific result is rated with one bullet if it is below the 20% of the worst value; two dots if it is between 20 and 40%, and so on. Then, we compute the mean for each FS method across all datasets. Notice that, since the error and distance measures are not the same for ranking and subset methods, we separate accordingly the methods to calculate the number of dots, for the sake of fairness.

According to the summary presented in Table 2.12, it remains clear the predominance of ReliefF in terms of accuracy of the selection, although InfoGain shows better performance according to stability. The methods that require a smaller training time in this set of experiments are mRMR and FCBF. Notice that, in this table, we are showing the results on some the artificial datasets presented in Sect. 2.3.1 (Corral, Led, Monk3, XOR and Parity3+3) as well as Monk1 and Monk2 (see [37] for more details).

Due to their particularities, SD1, SD2 and SD3 were evaluated separately. Table 2.13 provides some guidelines for the specific scalability aspects considered for the SD datasets. Focusing on the ranker methods, one can see that in terms of error, mRMR seems to be the best alternative. In turn, with regard to the distance, Chi-Squared and InfoGain –univariate methods– return stable rankings (low distance), while ReliefF and mRMR –multivariate methods– show high distances. This might seem surprising, since one can expect that a method which obtains the minimum error would also be highly stable. However, SD datasets have the particularity of consisting of groups of features equally relevant, but redundant among each other so when one of the features in the group is selected (or ranked top) the remaining ones in the group are deemed as irrelevant. Therefore, if we have two groups of relevant features (10 in each group), as it is the case with SD1 (see Sect. 2.3.1), mRMR can

Table 2.13 Overview of scalability of filters on SD datasets (notice that the large the number of dots, the better the behavior)

	Method	Error	Distance	Training time
Ranking	Chi-Squared	●●●	●●●●	●●●●●
	ReliefF	●●	●●●	●●●●●
	InfoGain	●●●	●●●●	●●●●●
	mRMR	●●●●	●●	●
Subset	FCBF	●●●	●●	●●●●●
	CFS	●●	●●	●
	Consistency	●●	●●	●●●●
	INTERACT	●●	●●●	●●●●

be selecting on the top of the ranking features 1 and 11 in the first repetition, 2 and 12 in the second repetition and so on. In this case, the error will be 0, but the method will achieve a high distance since the rankings are very dissimilar. Moreover, when performing several repetitions of a particular experiment, it is common that low minimum errors are obtained together with high variance. To sum up, in this scenario the best ranker in terms of error seems to be mRMR, but at the cost or requiring large amounts of time and not being stable. In turn, InfoGain and Chi-Square obtain also acceptable errors, they become stable although require certain amount of data, and the computational cost is acceptable (low training time).

The training time required by mRMR is in the order of thousands of seconds while the remaining methods require in the order of seconds. The only of these methods which theoretical complexity is quadratic to the number of features is mRMR, and thus its time raises significantly as the number of features increases (in these experiments, up to 4096 features). Notice that ReliefF in this case does not require high times as in the previous set of experiments, because it is quadratic in the number of samples and linear in the number of features, and SD datasets have a small number of samples.

In the case of subset filters, it is worth mentioning that the poor distance results are due to the fact that the SD datasets have an extremely high number of features (up to 4096), so the more features, the more difficult is to select stable subsets of features. Having said that, FCBF is clearly the best option since it obtained the best results in terms of error and training time, although at the cost of being the least stable.

Table 2.14 shows the scalability results of wrapper methods. In this case, the wrapper model was evaluated with three representative classifiers (C4.5, k-NN and naive Bayes) to assess the relative usefulness of the subsets of variables. Notice that the search strategy is *best first*, starting with the empty set of features and searching forward (which tends to select larger subsets of features than the backward search strategy). The best learning algorithm to be combined with the wrapper seems to be C4.5. Although k-NN shows a good behavior in terms of error and distance, the training time is quite high, probably because of the time burden required to sort the samples in order to find the nearest neighbors. C4.5 achieves a lower error and distance in a shorter time. Notice that studying the scalability of wrapper methods is not as straightforward as with filters, since the former involve learning algorithms, which add more complexity to the task.

Finally, we have also studied the scalability of two embedded methods: FS-P and SVM-RFE. Preliminary experiments on Corral dataset showed that the maximum training time required by SVM-RFE was almost 18000 s. This high time is due to the recursive nature of the method thus preventing its application to the remaining datasets, that are more complex than Corral. From the results on Corral dataset, it could be seen that in terms of error, the embedded methods were more affected by the number of samples than of features. Regarding the distance, their performance is comparable to those of the multivariate ranker filters ReliefF and mRMR, being more affected by the number of features than of samples, due to the possible combinations of features. Focusing on the training time, SVM-RFE required a much longer time than FS-P. For this reason, FS-P seemed to be a better option since in the rest of the measures the performance is similar. However, only one dataset is not enough to draw strong conclusions.

In light of the results presented in this section, some guidelines have been proposed:

- Among the subset filters, INTERACT obtained good scalability results, especially in terms of minimum error and distance. However, if we are interested in really low training times, FCBF showed an acceptable accurate results in a short training time.
- As for the ranker methods, ReliefF turned out to be very precise in selecting the relevant features, although this comes at the prize of large training times when the number of samples is high. On the other hand, the ranker method mRMR also achieved good results in terms of error, but the training time raises significantly when it deals with large amounts of features.
- With regard to the stability of the filters evaluated, i.e. the sensitivity of the methods to variations in the training set, univariate ranker methods (such as Chi-Squared or Information Gain) are more stable than multivariate methods (such as ReliefF or mRMR), since the latter have to deal with interactions between features. It is worth mentioning that subset methods, although being also multivariate, are much more stable than their counterparts within the ranker approach. This is happening

Table 2.14 Overview of the scalability of wrappers (notice that the large the number of dots, the better the behavior)

Method	Error	Distance	Training time
W-C45	••••	••	•••••
W-k-NN	••••	•••	•••
W-NB	•••	••	•••••

because ranker methods have to order all the features, even the irrelevant ones, while subset methods only select the relevant features so if they are behaving correctly, it is easier to select consistent subsets of features.

- As expected, the scalability of the wrapper methods depends on the classifier chosen, being C4.5 a good option in terms of scalability.
- When using embedded methods, the well-known SVM-RFE algorithm achieves promising results in terms of error at the cost of requiring high training times.
- In general, the authors suggest the use of filters, since they carry out the feature selection process with independence of the induction algorithm and are faster than embedded and wrapper methods, scaling better to Big Data problems.

2.5 Summary

In this chapter we have explained the foundations of feature selection, which will be the base learners of the ensemble models analyzed in this book. First, we started by defining the difference between feature selection and feature extraction and the different approaches for feature selection. Some state-of-the-art methods were briefly described and then reviewed through the analysis of their performance when facing a total of 11 synthetic datasets covering different situations such as presence of irrelevant and redundant features, noise in the data or interaction between attributes. A scenario with a small ratio between number of samples and features where most of the features are irrelevant was also tested. It reflects the problematic of datasets such as microarray data, a well-known and hard challenge in the machine learning field where feature selection becomes indispensable. And, finally, we have also provided a study about the scalability of these state-of-the-art methods, an issue that has not received much consideration in the literature. Again, the methods were evaluated facing a set of 11 artificial datasets.

References

1. Wold, S., Esbensen, K., Geladi, P.: Principal component analysis. Chemom. Intell. Lab. Syst. **2**(1–3), 37–52 (1987)
2. Yang, Y. Pederson, J.O.: A comparative study on feature selection in text categorization. In: Proceedings of the 20th International Conference on Machine Learning, pp. 856–863 (2003)
3. Yu, L., Liu, H.: Efficient feature selection via analysis of relevance and redundancy. J. Mach. Learn. Res. **5**, 1205–1224 (2004)
4. Provost, F.: Distributed data mining: scaling up and beyond. J. Adv. Distrib. Parallel Knowl. Discov. 3–27 (2000)
5. Guyon, I., Gunn, S., Nikravesh, M., Zadeh, L.: Feature Extraction: Foundations and Applications. Springer, Berlin (2006)
6. Stańczyk, U., Jain, L.C.: Feature Selection for Data and Pattern Recognition. Springer (2015)
7. Liu, H., Motoda, H.: Computational Methods of Feature Selection. CRC Press (2007)

8. Bolón-Canedo, V., Sánchez-Maroño, N., Alonso-Betanzos, A.: Feature Selection for High-dimensional Data. Springer (2015)
9. Hall, M.A.: Correlation-based Feature Selection for Machine Learning. Ph.D. thesis, University of Waikato, Hamilton, New Zealand (1999)
10. Dash, M., Liu, H.: Consistency-based search in feature selection. J. Artif. Intell. **151**(1–2), 155–176 (2003)
11. Zhao, Z., Liu, H.: Searching for interacting features. In: Proceedings of the International Joint Conference on Artificial Intelligence, pp. 1156–1167 (1991)
12. Hall, M.A., Smith, L.A.: Practical feature subset selection for machine learning. J. Comput. Sci. **98**, 4–6 (1998)
13. Kononenko, I.: Estimating attributes: analysis and extensions of RELIEF. In: Proceedings of the European Conference on Machine Learning, pp. 171–182 (1994)
14. Kira, K., Rendell, L.: A practical approach to feature selection. In: Proceedings of the 9th International Workshop on Machine Learning, pp. 249–256 (1992)
15. Peng, H., Long, F., Ding, C.: Feature selection based on mutual information: criteria of max-dependency, max-relevance, and min-redundancy. IEEE Trans. Pattern Anal. Mach. Intell. **27**(8), 1226–1238 (2005)
16. Guyon, I., Weston, J., Barnhill, S.M.D., Vapnik, V.: Gene selection for cancer classification using support vector machines. J. Mach. Learn. **46**(1–3), 389–422 (2002)
17. Rakotomamonjy, A.: Variable selection using SVM-based criteria. J. Mach. Learn. Res. **3**, 1357–1370 (2003)
18. Mejía-Lavalle, M., Sucar, E., Arroyo, G.: Feature selection with a perceptron neural net. In: Proceedings of the International Workshop on Feature Selection for Data Mining, pp. 131–135 (2006)
19. Witten, I.H., Frank, E.: Data mining: practical machine learning tools and techniques. Morgan Kaufmann, San Francisco. http://www.cs.waikato.ac.nz/ml/weka/ (2005). Accessed July 2017]
20. Belanche, L.A., González, F.F.: Review and evaluation of feature selection algorithms in synthetic problems. http://arxiv.org/abs/1101.2320. Accessed July 2017
21. John, G.H., Kohavi, R., Pfleger, K.: Irrelevant features and the subset selection problem. In: Proceedings of the 11th International Conference on Machine Learning, pp. 121–129 (1994)
22. Zhu, Z., Ong, Y.S., Zurada, J.M.: Identification of full and partial class relevant genes. IEEE Trans. Comput. Biol. Bioinform. **7**(2), 263–277 (2010)
23. Thrun, S. et al., The MONK's problems: A performance comparison of different learning algorithms. Technical report CS-91-197, CMU (1991)
24. Breiman, L., Friedman, J., Olshen, R., Stone, C.: Classification and Regression Trees. Wadsworth International Group (1984)
25. Mamitsuka, H.: Query-learning-based iterative feature-subset selection for learning from high-dimensional data sets. Knowl. Inf. Syst. **9**(1), 91–108 (2006)
26. Quinlan, J.R.: C4.5: Programs for Machine Learning. Morgan Kaufmann (1993)
27. Rish, I.: An empirical study of the naive bayes classifier. In: Proceedings of IJCAI-01 Workshop on Empirical Methods in Artificial Intelligence, pp. 41–46 (2001)
28. Aha, D.W., Kibler, D., Albert, M.K.: Instance-based learning algorithms. J. Mach. Learn. **6**(1), 37–66 (1991)
29. Shawe-Taylor, J., Cristianini, N.: An Introduction To Support Vector Machines And Other Kernel-based Learning Methods, Cambridge University Press (2000)
30. Bolon-Canedo, V., Sanchez-Marono, N., Alonso-Betanzos, A.: A review of feature selection methods on synthetic data. Knowl. Inf. Syst. **34**(3), 483–519 (2013)
31. Kohavi, R., John, G.H.: Wrappers for feature subset selection. J. Artif. Intell. **97**(1–2), 273–324 (1997)
32. Kim, G., Kim, Y., Lim, H., Kim, H.: An MLP-based feature subset selection for HIV-1 protease cleavage site analysis. J. Artif. Intell. Med. **48**, 83–89 (2010)
33. Seijo-Pardo, B., Bolón-Canedo, V., Alonso-Betanzos, A.: Testing different ensemble configurations for feature selection. Neural Process. Lett. **46**, 857–880 (2017)

34. Bolón-Canedo, V., Sánchez-Maroño, N., Alonso-Betanzos, A.: Recent advances and emerging challenges of feature selection in the context of big data. Knowl.-Based Syst **86**, 33–45 (2015)
35. Khoshgoftaar, T M., Golawala, M. and Van Hulse, J. An empirical study of learning from imbalanced data using random forest. In: ICTAI 2007. 19th IEEE International Conference on Tools with Artificial Intelligence, vol. 2, pp. 310–317. IEEE (2007)
36. Liu, H. and Setiono, R.Chi2: Feature selection and discretization of numeric attributes. In: Proceedings of Seventh International Conference on Tools with Artificial Intelligence, pp. 388–391. IEEE (1995)
37. Bolón-Canedo, V., Rego-Fernández, D., Peteiro-Barral, D., Alonso-Betanzos, A., Guijarro-Berdiñas, B., Sánchez-Maroño, N.: On the scalability of feature selection methods on high-dimensional data. Knowl. Inf. Syst. (2017, in press)

Chapter 3
Foundations of Ensemble Learning

Abstract This chapter describes the basic ideas under the ensemble approach, together with the classical methods that have being used in the field of Machine Learning. Section 3.1 states the rationale under the approach, while in Sect. 3.2 the most popular methods are briefly described. Finally, Sect. 3.3 summarizes and discusses the contents of this chapter.

The idea of combining multiple models instead of a single model to solve a given problem has its rationale in the old proverb "Two heads are better than one". The approach constructs a set of hypothesis using several different models, that then are combined in order to be able to obtain better performance than learning just one hypothesis using a unique method [1–3]. There have been several studies that have shown that these models obtain usually better accuracy than individual methods, due to the diversity of the approaches and the control of the variance [3]. These combinations of models are called "committees", or more recently "ensembles". Ensemble learning algorithms have reached great popularity among the Machine Learning literature, as they achieve performances that were not possible some years ago, and thus have become a "winning horse" in many applications [4].

3.1 The Rationale of the Approach

As early as 1785, the Condorcet's jury theorem established what can be considered the initial seed of the modern ensembles, that is, that a correct decision in a problem could be obtained combining the individual votes of a large enough jury, given some restrictions. This idea is the rationale under ensemble learning. The contributions of the field of Machine Learning to combining classifiers started with Tukey's work [5] in the 1970's, but in the 1990's the interest in the field increased, and many works, and the classical ensemble approaches, date from that decade [6–8]. There are several reasons [6, 9] for the ensembles to be able to improve accuracy over single methods, that can be summarized as follows: The "No Free Lunch" theorems have shown that learning algorithms can not be universally good, and thus no algorithm

© Springer International Publishing AG, part of Springer Nature 2018   39
V. Bolón-Canedo and A. Alonso-Betanzos, *Recent Advances in Ensembles*
for Feature Selection, Intelligent Systems Reference Library 147,
https://doi.org/10.1007/978-3-319-90080-3_3

can outperform any other algorithm when performance is uniformly averaged over all target functions. With the myriad of algorithms available, it is almost impossible for a user to find the best algorithm for each given problem, and thus the advantage of ensembles is that by combining multiple models, a "meta" learning scheme is build, producing an output that combines different hypothesis spaces, and hopefully will help to obtain approaches at least "near" the ideal solution for most cases. Of course, when using an ensemble we avoid choosing a single method, but a new issue appears, that is, which methods are the best for being included in the ensemble, and how many of them are necessary. In fact, this is a Multiple Criteria Decision Making (MCDM) problem, for which several techniques could be used in turn in order to choose the optimal options. In the extensive review work in the subject of MCDM carried out by Mardani et al. [10], they state that the Analytic Hierarchy Process (AHP) method [11] and the hybrid MCDM method [12–14] are the most common in use nowadays. Some of the most employed selection criteria to be combined are the following:

- Accuracy of each of the components and of the ensemble.
- Cost, the computational complexity and time needed by the methods.
- Diversity which should be encouraged in the methods that form the ensemble, to obtain an adequate mixture of hypotheses. In [1, 15] several methods for creating diversity are described in detail.
- Parameter optimization, the set of parameters to be controlled should be comprehensive and easy to tune, so usability of the final ensemble should be guaranteed.
- Scalability, the ensemble methods should assure scalability to large datasets, ideally in both samples and features. This is related to the cost aspect, since most algorithms have a computational complexity greater than linear (in samples, features or both), and also large datasets increase the size of the search space, and thus the problem is more difficult for machine learning algorithms.

Ensembles have been used extensively in supervised classification, regression and optimization, as the combination of these was the first natural step in the field of machine learning, and has received considerable attention during years [6, 16, 17]. *Bagging* [18] and *Boosting* [19], are the classical ensemble classification methods, that have been developed with the objective of reducing overfitting, while improving accuracy, and thus have been widely used by researchers. However, there are also another areas of Machine Learning that have used the approach, such as clustering, in which ensembles have been used for improving the robustness, novelty and stability of unsupervised learning solutions [20–23], or even to pursue scalable approaches to the problem [24, 25]. One recent area in which ensembles have demonstrated their interest is quantification [26], as its aim is to estimate the number of cases that belongs to each class in a test set, using a training set that perhaps has a different distribution. Although intuitively counting the predictions of a classifier will do the work, this last approach is not appropriate, as it does not take into account the differences in distributions between training and testing datasets. However, ensembles philosophy fits adequately with this distribution shift. Feature selection is another task in which the ensemble philosophy has encountered place, basically for the same reasons, that is robustness, diversity and scalability, but also trying to overcome the variability of results that the different feature selection methods available obtain for different datasets, as there is not a method that works universally well. Thus, in order to select the most adequate method, the users should have an in-depth knowledge of all the strengths and weaknesses of individual methods. Using an ensemble allows for relieving the user from that decision, while obtaining usually best performances than individual methods alone [27–32].

3.2 Most Popular Methods

There are several aspects of the ensembles that can be used to classify different types of ensembles. The main blocks of an ensemble are the training set, the base inducer, and the combiner method used to aggregate the individual results obtained. The training set is the labelled data set that is used to train the model, it can be the same for all base classifiers, or a different one, for example using different partitions of the original dataset. The base inducer is the induction algorithm used that obtains an individual result (a classification, a feature subset, etc.) for the problem. The base methods can be generated using the same generation algorithm or a different one. Also, even if the algorithm is the same, different parameters could also be used in each base classifier, giving raise to more diversity [15, 33]. Finally, the results of the individual base methods should be aggregated to obtain a final decision, there are several aggregation methods than can be used, ranging from the simple mean, median or majority voting to more sophisticated algorithms. For example, in the case of classification problems, majority voting is the most popular combination method chosen, while for regression averaging is the predominant strategy. Besides, the individual models of the ensemble might be treated equally or they can be weighted according to some rule that usually depends on the accuracy of the results obtained [6, 34] (this issue will be further developed in Chap. 5).

Thus, attending to the base methods that form them, we could differentiate two types of ensembles [16]:

- Homogeneous, when all base classifiers are of the same type (see Fig. 3.1)
- Heterogeneous, if the base classifiers are not of the same kind (see Fig. 3.2).

These two basic types could be even more differentiated, depending on the combination of the following factors, giving rise to a number of different ensemble configurations:

- If the base methods have been generated using the same generation algorithm or not.
- If the base methods are generated using the same generation algorithm, but using different parameter settings in each one.

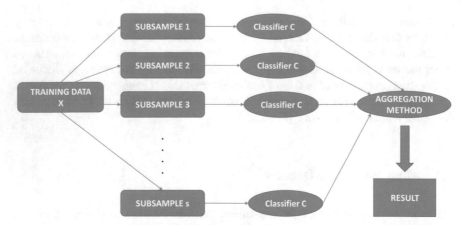

Fig. 3.1 A scheme of an homogeneous ensemble of classifiers. As can be seen, the same classification method (C) is applied over different subsets generated from the training set

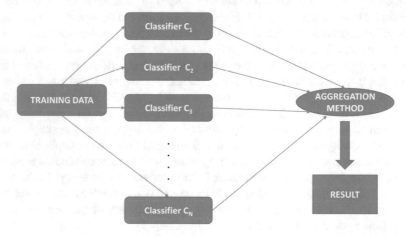

Fig. 3.2 A scheme of an heterogeneous distributed ensemble. As can be seen, a different classification method (C_i) is applied over the training set, and the different results are to be aggregated, for example using majority voting, for obtaining a final result

- If the training dataset is the same or not for all base methods.
- If the attributes of the training dataset are the same or not for all base methods.

In the case of ensembles for feature selection, and although they will be described in detail in Chap. 4, an example of a scheme for both, homogeneous and heterogeneous approaches is given below, in Figs. 3.3 and 3.4, for an easier comparison with the same schemes for the classifiers.

On the following we will describe the most common and classical ensemble methods for classification, as this is the area of data mining in which ensembles appeared. First, we will describe Bagging and Boosting, two of the most popular models, as

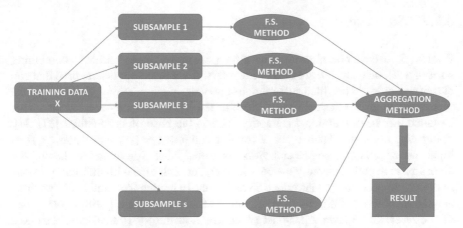

Fig. 3.3 A scheme of an homogeneous ensemble of feature selectors, in which the same feature selection algorithm is applied over different subsets derived from the training set

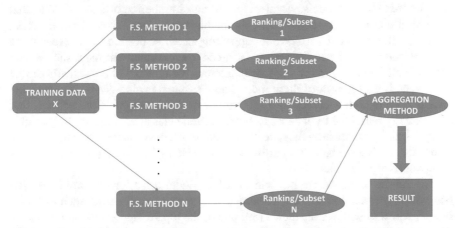

Fig. 3.4 A scheme of an heterogeneous distributed ensemble, in which different feature selection algorithms are applied to a training set, and then the results (that might be a subset or a rank, need to be aggregated to obtain a final feature selection result

both are based on introducing diversity by modifying the training set, in such a way that the learning algorithm is executed multiple times over different training sets. Then, the also very well-known Random Forest approach is also described, as it is one of the most used classifiers, due to their good results in performance. In fact, in [35], they were found to be the best over a testbed of 179 classifiers belonging to different families. Although in [36], these results are criticized for being biased by the lack of a held-out test set and the exclusion of trials with errors, still Random Forest is among the best performing methods, together with Support Vector Machines (SVM) and Neural Networks (NN).

3.2.1   Boosting

Boosting is an ensemble algorithm that aims at creating a strong learner by combining a set of weak ones, which are the base learners. A weak learner is a method that can classify samples better than just random guessing, while a strong classifier is one that achieves error rates arbitrarily close to the irreducible Bayes error rate. This is called the weak learning assumption, and is the base of the method [37, 38]. A boosting algorithm takes a set of training samples $I = ((x_1, y_1), ...(x_n, y_n))$ as input, where x_i is an instance and y_i its associated label. The main idea behind the algorithm is to call repeatedly the base learner, but each time with different portions of the training set, so as to introduce diversity, until a predefined number of iterations is reached. There are different variants of the basic Boosting algorithm, being one of the most well-known AdaBoost (Adaptive Boosting) [8]. AdaBoost assigns a weight to each instance of the data set, that initially is the same for all patterns. These weights are iteratively updated, attempting to give more importance to the samples that have been misclassified, and decreasing the importance of those that have been classified correctly. Thus, the higher the weight for an input instance, the higher the chance it will have for appearing in a new portion sample to call the base method again. Thus, the learner is forced to concentrate on the instances that are more difficult, by performing more iterations with them and thus creating more classifiers. Besides, each of these individual classifiers is also given a weight, that measures its overall accuracy as a function of the total weight of the patterns that are correctly classified by it. That is, more accurate classifiers have higher weights assigned, and consequently they are more used for the classification of new patterns. Thus, the AdaBoost algorithm performs an iterative procedure that uses a series of complementary classifiers.

The AdaBoost algorithm has win the 2003 Gödel prize for its authors, having been demonstrated that it can approximate a large margin classifier, such as SVM, being according with some studies [39] one of the top 10 algorithms in the field of Machine Learning. In Algorithm 3.1, the pseudocode for the basic strategy of the AdaBoost algorithm is shown [3], for the case of a binary classification, that is when $y_i \in \{-1, +1\}$. Some previous nomenclature is needed; supposing that we have N samples (instances), the weight loss of AdaBoost with the strong learner H is of exponential type, as in that way loss is higher and abrupt when the predictions of the classifier are wrong. Let $\mathscr{L}(H)$ be the weight loss of Adaboost with the strong learner H such that,

$$\mathscr{L}(H) = \sum_{i=1}^{N} e^{-m_i}$$

where m_i is the voting margin for each of the k base weak learners, and can be calculated as:

$$m_i = y_i \sum_{k=1}^{k} \alpha_k h_k(x_i),$$ where α_k is the weight of the k learner and $h_x(x_i)$ is the prediction over x_i of the k learner.

Algorithm 3.1: Basic Pseudo-code for AdaBoost

Data: $I = ((x_1, y_1), ...(x_n, y_n))$ = training dataset where $y_i \in \{-1, +1\}$; the size (K) of the ensemble; the weak learners

1   $\forall_i D_0(i) \leftarrow (\frac{1}{N})$, Set uniform weights for the samples of the training set

2   **for** $k = 1$ *to* k; *that is for each base learner of the ensemble* **do**

3      $h_k \leftarrow$ base learner trained on D; Train a model h_k using distribution D_k

4   Calculate the error; $\varepsilon_k \leftarrow \sum_{i=1}^{N} D_{k-1}(i)$

5   **if** $\varepsilon_k \geq 0.5$ **then**

6 | exit loop

     end

7   Set $\alpha_k = \frac{1}{2}ln\frac{1-\varepsilon_k}{\varepsilon_k}$; Set sample weights based on ensemble predictions

8   Update $D_{k+1}(i) = \frac{D_k(i)exp(-\alpha_k y_i h_k(x_i))}{Z_k}$, being Z_t a normalization factor so to make D_{k+1} a valid distribution

   end

9   **for** *a new testing point* (x', y') **do**

10      $H(x') = sign(\sum_{k=1}^{K} \alpha_k h_k(x'))$

   end

In general, AdaBoost improves accuracy, although sometimes it might fail due to overfitting [40], as a large number of iterations might produce an over-complex learner. Besides, in the present context of Big Data, AdaBoost as well as other boosting algorithms, presents several problems on computing complexity and poor learning accuracy, as large datasets increase the size of the search space, and thus the possibility of selecting an overfitted learner. There have been several attempts to improve the results of AdaBoost over large datasets. The P-AdaBoost is a par-allellized version [41] which builds upon earlier results concerning the dynamics of AdaBoost weights. Boosting-by-Resampling [42] uses a local error measure to avoid the negative contribution of noisy samples, more common in large datasets. Boosting can also be viewed as a functional gradient descent technique, and that was the view of the work in [43], in which the author combined it with the L_2-loss function to develop a computational variant of Boosting named L_2Boost. In [44] Boosting was applied over a high dimensional dataset on gene expression data in conjunction with decision trees, in order to achieve a pre-selection of the variables involved. Finally, Buhlman [45] proved that boosting with that squared error loss, L_2Boost, is consistent under certain assumptions, for very high-dimensional linear models, where the number of predictor variables is allowed to grow as fast as the sample size. The author also proposes a method that allows for choosing the number of boosting iterations, thus making the algorithm computationally interesting, as it is not required to run it multiple times for cross-validation, as it has been the common procedure so far.

Boosting has also been adapted for cost-sensitive learning, in which the false positive or false negative predictions or classifications should be weighted differently, normally due to the characteristics of the problem at hand. For example, a false negative prediction of an illness might be a bigger problem than a false positive, as a real illness is undetected, while a false positive in classifying an attack in a computer network can results in an unnecessary Denial-of-Service. There are several variants of boosting algorithms that deal with this problem [46–48]. However, in an interesting article [49], the authors demonstrate that cost-sensitive modifications seem unnecessary for Adaboost, if proper calibration is applied.

3.2.2 Bagging

Bagging (Bootstrap AGGregatING) is another popular ensemble method. It works by creating diverse models on different random samples of the original dataset, and then combine these models to obtain a final result. This strategy is effective when the learners are unstable and tend to be reactive to little variations within the input space, as in the case of neural networks or decision trees. The samples, known as bootstrap samples, are taken uniformly and with replacement. Thus, due to the method employed, samples may usually contain duplicates, and also perhaps some of the original data instances could be missing, even if both, the bootstrap sample and the original dataset are of the same size [50]. Specifically, if we have N samples, the bootstrap procedure samples uniformly with replacement, being the probability of an individual instance not being selected $(1 - \frac{1}{n})^N$, and consequently with a large N, a single bootstrap is expected to contain approximately 63.2% of the original set, while 36.8% of the original sample items are not selected [3]. The training sets are different one from another, but they are not statistically independent. These differences in the bootstrap samples are aimed at creating diversity among the models of the ensemble.

The pseudocode of bagging is shown in Algorithm 3.2

Algorithm 3.2: Basic Pseudo-code for Bagging

Data: I = $((x_1, y_1), ...(x_n, y_n))$ = training dataset where $y_i \in \{-1, +1\}$; the size (K) of the
ensemble; the weak learners

1 **for** $k = 1$ *to K; that is for each base learner of the ensemble* **do**
2   Build a bootstrap sample I_t from I by sampling $|I|$ data points with replacement
3   Run the weak learner in I_t to produce a model M_t
  end
4 Return $\{M_t|1\}$

We can see that the bagging ensemble returns a set of models, which results should be combined. Normally, this is done by majority voting or by averaging. As a result, bagging produces a combined model that usually performs better than the single model trained with the original data, and as boosting, it does so using the same inducer method.

A special and very common type of bagging is Random Forest (RF), in which bagging is used in combination with tree models, and also the trees are built from a different random subset of the features (subspace sampling). This raises even more the diversity of the ensemble, and besides reduces the training time of each tree. The Random Forest algorithm (3.3 shows the corresponding pseudocode), is a very popular ensemble learning method for classification, regression and other tasks, that constructs a multitude of decision trees at training time and that outputs the class that is the mode of the classes (for classification), or the mean prediction (for regression) of the individual trees. The main advantage of RF is that they can greatly reduce or even avoid overfitting by optimizing a tuning parameter that governs the number of features that are randomly chosen to grow each tree from the bootstrapped data. Typically, this is carried out employing a k-fold cross-validation (with k usually in the interval [5, 10], with both extreme values as the most common), choosing the tuning parameter that minimizes test sample prediction error. Besides, growing a larger forest will improve predictive accuracy, although there are usually diminishing returns once certain sizes are achieved (in the order of several hundreds).

RF were first proposed in Ho's work [51], trying to overcome the problems of traditional tree methods that cannot be grown to arbitrary complexity as there is the risk of possible loss of generalization accuracy on unseen data. This limitation on complexity usually means suboptimal accuracy on training data. However, decision trees are of interest in Machine Learning due to their high execution speed, thus Ho's work proposed a method to build tree-based classifiers, in randomly selected subspaces of the feature space, which capacity can be arbitrarily expanded for increases in accuracy for both training and unseen data. Later, Breiman's work [52] introduced bootstrapping aggregation for the independence of each base classifier, that is theoretically enforced by training each decision tree on a training set sampled with replacement from the original training set (bagging). Besides, further randomness is introduced by identifying the best split feature from a random subset of available features. The ensemble classifier then aggregates the individual predictions to combine into a final prediction, based on a majority voting on the individual predictions (see Fig. 3.5). In this way, bias and variance both can be reduced, thus making the model more robust and accurate.

Bagging and Boosting are two different sides of the bias-variance dilemma (Fig. 3.6). This dilemma states than a low complexity model suffers less from variability due to random variations in the training data, although it may introduce a systematic bias that even a large amount of training data can not solve. On the other side, high-complexity models eliminate the bias, but may suffer non systematic errors due to variance [53]. While boosting is mainly a bias reduction technique, bagging is a variance reduction technique. For that reason, bagging is often combined with high-variance models such as trees, while boosting is combined more frequently with high-bias models such as linear classifiers or univariate decision trees.

Algorithm 3.3: Basic Pseudo-code for Random Forest

Data: I = $((x_1, y_1), ...(x_n, y_n))$ = training dataset where $y_i \in \{-1, +1\}$; the size (K) of the ensemble; subspace dimension d

1 for $k = 1$ *to K; that is for each base learner of the ensemble* **do**
2 │ Build a bootstrap sample I_t from I by sampling $|I|$ data points with replacement
3 │ Select d features at random diminishing the dimensionality of I_t
4 │ Train a tree model M_t on I_t without pruning
 end
5 Return $\{M_t | 1\}$

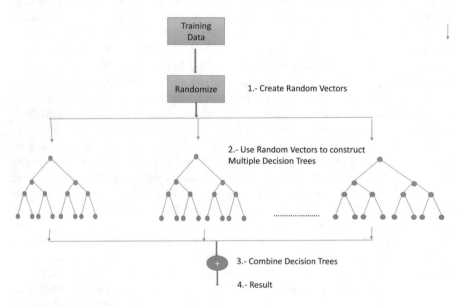

Fig. 3.5 A scheme of random forest

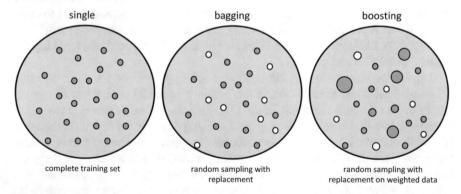

Fig. 3.6 Example of data sampling in the case of a single learner, and bagging and boosting approaches

Bagging and Boosting use the same learner model for the ensemble, but as stated in the previous section, it is also possible to combine different based models into an ensemble, that is named heterogeneous ensemble. In this case, the diversity comes from the fact that the base learners are trained by different learning algorithms, and normally they use the same training set in this scheme. Another possibility is to use the same base models, but employing different parameter settings, as mentioned before. In all these cases, combination methods are to be used, and these will be further detailed in Chap. 5.

Bagging and Boosting algorithms, including Random Forest, are available in several of the most common Machine Learning frameworks, such as Scikit Learn,[1] MatLab,[2] or Weka.[3] For more detailed information, please consult Chap. 9.

3.3   Summary

An ensemble is formed by a set of base models and a meta-model that is trained to decide how to combine the base models learning. If the meta-model is linear, it is known as stacking. The ensemble is formed by those base learners, and they might be the same or different models, using the same or different parameters, and the same or different training sets, thus giving an ample repertory of possible models. The rationale over the use of ensembles for learning is that by using several models, taking into account that diversity must be assured, better results in accuracy will be obtained. The results of the individual models of the ensembles must be combined so as to obtain a final model. Several strategies could be used, being the most common the majority voting or the average.

References

1. Rokach, L.: Ensemble-based classifiers. Artif. Intell. Rev. **33**(1), 1–39 (2010)
2. Polikar, R.: Ensemble based systems in decision making. IEEE Circuits Syst. Mag. **6**(3), 21–45 (2006)
3. Brown, G.: Ensemble learning. In: Sammut, C., Webb, G.I. (eds.) Encyclopedia of Machine Learning. Springer, Berlin (2010)
4. Caruana, R., Niculescu-Mizil, A.: An empirical comparison of supervised learning algorithms. In: Machine Learning: Proceedings of the 23rd International Conference, pp. 161–168. ACM (2006)
5. Tukey, J.W.: Exploratory Data Analysis. Addison-Wesley, Reading (1977)
6. Kuncheva, L.I.: Combining Pattern Classifiers: Methods and Algorithms. Wiley, New York (2004)
7. Freund, Y.: Boosting a weak learning algorithm by majority. Inf. Comput. **121**(82), 256–285 (1995)

---

[1] http://scikit-learn.org/stable/index.html.

[2] https://es.mathworks.com/help/stats/classification-ensembles.html?requestedDomain=true.

[3] https://machinelearningmastery.com/use-ensemble-machine-learning-algorithms-weka/.

8. Freund, Y., Schapire, R.E.: Experiments with a new boosting algorithm. In: Machine Learning: Proceedings of the Thirteenth International Conference, pp. 325–332 (1996)
9. Dietterich, T.G., Bakiri, G.: Solving multiclass learning problems via error-correcting output codes. J. Artif. Intell. Res. **2**, 263–286 (1995)
10. Mardani, A., Jusoh, A., Nor, K., Khalifah, Z., Zakwan, N., Valipour, A.: Multiple criteria decision-making techniques and their applications -a review of the literature from 2000 to 2014. Econ. Res. Ekon. Istraivanja **28**(1), 516–571 (2015)
11. Saaty, T.: What is the analytic hierarchy process? In: Mitra, G., Greenberg, H., Lootsma, F., Rijkaert, M., Zimmermann, H. (eds.) Mathematical Models for Decision Support, vol. 48, pp. 109–121. Springer, Berlin (1988)
12. Ilangkumaran, M., Karthikeyan, M., Ramachandran, T., Boopathiraja, M., Kirubakaran, B.: Risk analysis and warning rate of hot environment for foundry industry using hybrid MCDM technique. Saf. Sci. **72**, 133–143 (2015)
13. Hashemkhani Zolfani, S., Esfahani, M.H., Bitarafan, M., Zavadskas, E.K., Arefi, S.L.: Developing a new hybrid MCDM method for selection of the optimal alternative of mechanical longitudinal ventilation of tunnel pollutants during automobile accidents. Transport **28**, 89–96 (2013)
14. Hu, S.-K., Lu, M.-T., Tzeng, G.-H.: Exploring smart phone improvements based on a hybrid MCDM model. Expert Syst. Appl. **41**, 4401–4413 (2014)
15. Brown, G., Wyatt, J., Harris, R., Yao, X.: Diversity creation methods: a survey and categorization. Inf. Fusion **6**(1), 5–20 (2005)
16. Bramer, M.: Principles of Data Mining. Springer, Berlin (2013)
17. Bishop, C.M.: Pattern Recognition and Machine Learning. Springer, Berlin (2006)
18. Breiman, L.: Bagging predictors. Mach. Learn. **24**(2), 123–140 (1996)
19. Schapire, R.E.: The strength of weak learnability. Mach. Learn. **5**(2), 197–227 (1990)
20. Vega-Pons, S., Ruiz-Shulcloper, J.: A survey of clustering ensemble algorithms. Int. J. Pattern Recognit. Artif. Intell. **25**(3), 337–372 (2011)
21. Hu, J., Li, T., Luo, C., Fujita, H., Yang, Y.: Incremental fuzzy cluster ensemble learning based on rough set theory. Knowl. Based Syst. **132**, 144–155 (2017). https://doi.org/10.1016/j.knosys.2017.06.020
22. Dimitriadou, E., Weingessel, A., Hornik, K.: A cluster ensembles framework. Design and Application of Hybrid Intelligent Systems. IOS Press, Amsterdam (2003)
23. Strehl, A., Ghosh, J.: Cluster ensembles—a knowledge reuse framework for combining multiple partitions. J. Mach. Learn. Res. **3**, 583–617 (2002)
24. Hore, P., Hall, L.O., Goldgof, D.B.: A scalable framework for cluster ensembles. Pattern Recognit. **42**(5), 676–688 (2009)
25. Hore, P., Hall, L., Goldgof, D.: A cluster ensemble framework for large data sets. In: Proceedings of IEEE International Conference on Systems, Man and Cybernetics, SMC'06, vol. 4, pp. 3342–3347 (2006)
26. Pérez-Gállego, P., Quevedo, J.R., del Coz, J.J.: Using ensembles for problems with characterizable changes in data distribution: a case study on quantification. Inf. Fusion **34**, 87–100 (2017)
27. Windeatt, T., Duangsoithong, R., Smith, R.: Embedded feature ranking for ensemble MLP classifiers. IEEE Trans. Neural Netw. **2286**, 988–994 (2011)
28. Attik, M.: Using ensemble feature selection approach in selecting subset with relevant features. In: Advances in Neural Networks-ISNN, pp. 1359–1366. Springer (2006)
29. Saeys, Y., Abeel, T., Van de Peer, Y.: Robust feature selection using ensemble feature selection techniques. Machine Learning and Knowledge Discovery in Databases. Springer, Berlin (2008)
30. Bolón-Canedo, V., Sánchez-Maroño, N., Alonso-Betanzos, A.: An ensemble of filters and classifiers for microarray data classification. Pattern Recognit. **45**(1), 531–539 (2012)
31. Yang, F., Mao, K.Z.: Robust feature selection for microarray data based on multicriterion fusion. IEEE/ACM Trans. Comput. Biol. Bioinform. (TCBB) **8**(4), 1080–1092 (2011)
32. Seijo-Pardo, B., Porto-Díaz, I., Bolón-Canedo, V., Alonso-Betanzos, A.: Ensemble feature selection: homogeneous and heterogeneous approaches. Knowl. Based Syst. https://doi.org/10.1016/j.knosys.2016.11.017

33. Kuncheva, L., Whitaker, C.: Measures of diversity in classifier ensembles and their relationship with the ensemble accuracy. Mach. Learn. **51**, 181–207 (2003)
34. Zhihua, Z.: Ensemble Methods: Foundations and Algorithms. Chapman and Hall/CRC. CRC Press, Boca Raton (2012)
35. Fernández-Delgado, M., Cernadas, E., Barro, S., Amorim, D.: Do we need hundreds of classifiers to solve real world classification problems? J. Mach. Learn. Res. **15**, 3133–3181 (2014)
36. Wainberg, M., Alipanahi, B., Frey, B.J.: Are random forests truly the best classifiers? J. Mach. Learn. Res. **17**, 1–5 (2016)
37. Schapire, R.E., Freund, Y.: Boosting. Foundations and Algorithms. The MIT Press, Cambridge (2012)
38. Schapire, R.E.: The strength of weak learnability. Mach. Learn. **5**, 197–227 (1990)
39. Wu, X., Kumar, V., Quinlan, J.R., Ghosh, J., Yang, Q., Motoda, H., et al.: Top 10 algorithms in data mining. Knowl. Inf. Syst. **14**(1), 1–37 (2008)
40. Quinlan, J.R.: Bagging, Boosting and C4.5. In: Proceedings of the 13th National Conference on Artificial Intelligence, pp. 725–730 (1996)
41. Merler, S., Caprile, B., Furlanello, C.: Parallelizing AdaBoost by weights dynamics. Comput. Stat. Data Anal. **51**(5), 2487–2498 (2007)
42. Zhang, C.X., Zhang, J.S.: A local boosting algorithm for solving classification problems. Comput. Stat. Data Anal. **52**(4), 1928–1941 (2008)
43. Buhlmann, P., Yu, B.: Boosting With the L2 Loss. J. Am. Stat. Assoc. **98**(462), 324–339 (2003)
44. Dettling, M., Bühlmann, P.: Boosting for tumor classification with gene expression data. Bioinformatics **19**(9), 1061–1069 (2003)
45. Buhlmann, P.: Boosting for high-dimensional linear models. Ann. Stat. **34**(2), 559–583 (2006)
46. Landesa-Vzquez, I., Alba-Castro, J.L.: Double-base asymmetric AdaBoost. Neurocomputing **118**, 101–114 (2013)
47. Ting, K. M.: A comparative study of cost-sensitive boosting algorithms. In: ICML, pp. 983–990 (2000)
48. Viola, P., Jones, M.: Fast and robust classification using asymmetric AdaBoost and a detector cascade. In: NIPS (2002)
49. Nikolaou, N., Edakunni, N., Kull, M., Flatch, P., Brown, G.: Cost-sensitive boosting algorithms: do we really need them? Mach. Learn. **104**(2–3), 359–384 (2016)
50. Breiman, M.: Bagging predictors. Mach. Learn. **24**(2), 123–140 (1996)
51. Ho, T.K.: Random decision forests. In: Proceedings of 3rd International Conference on Document Analysis and Recognition, vol. 1, pp. 278–282 (1995)
52. Breiman, L.: Random forests. Mach. Learn. **45**(1), 5–32 (2001)
53. Flach, P.: Machine Learning. The Art and Science of Algorithms that Make Sense of Data. Cambridge University Press, Cambridge (2012)
54. Bolón-Canedo, V., Sánchez-Maroño, N., Alonso-Betanzos, A.: Data classification using an ensemble of filters. Neurocomputing **135**, 13–20 (2014)

Chapter 4
Ensembles for Feature Selection

Abstract This chapter describes the ideas of the ensemble approach applied to feature selection, a classical preprocessing step which in the present context of Big Data and high dimensional datasets, has become of capital importance. Section 4.1 introduces the context of ensembles for feature selection, that are more detailed in Sects. 4.2 and 4.3 for homogeneous and heterogeneous ensembles, respectively. In both sections, a use case using rankers is employed to illustrate the concepts, in Sects. 4.2.1 and 4.3.1. Finally, in Sect. 4.4, a brief comparison between the results obtained by both approaches employed in the use cases is shown, with the aim of giving the readers a brief guideline of their better use.

In Chap. 3 we have seen the rationale under the ensemble approach for learning, specifically for classification and prediction. This same idea has been recently extended to other machine learning fields, such as quantification or feature selection. In this chapter the ideas of ensembles applied to feature selection will be detailed. Again, the basic assumption is the same, that is, that combining the outputs of several single feature selection models will obtain better results than using a single feature selection approach. Taking into account the different types of ensembles that were described in Chap. 3, we will focus on two different approaches: (i) homogeneous, that is, using the same feature selection method with different training data and distributing the dataset over several nodes (or several partitions); and (ii) heterogeneous, i.e., using different feature selection methods with the same training data. As in the general ensemble case, the results of the base selectors are to be combined to obtain a final result, and thus several aggregation methods can be used. In Sects. 4.2 and 4.3 both approaches are described in general terms, together with use cases of both approaches using rankers over a suite of datasets.

Part of the content of this chapter was previously published in *Knowledge-Based Systems* (https://doi.org/10.1016/j.knosys.2016.11.017).

© Springer International Publishing AG, part of Springer Nature 2018
V. Bolón-Canedo and A. Alonso-Betanzos, *Recent Advances in Ensembles for Feature Selection*, Intelligent Systems Reference Library 147,
https://doi.org/10.1007/978-3-319-90080-3_4

4.1 Introduction

In the last years data has been increasing in size (samples) and dimension (features) at unprecedented rates, due to the digitalization of most activities in several areas, as sensors are available for almost any task that can be think of. When data is used for machine learning, most of the times a preprocessing should be carried out, so as to eliminate noise, discretize data, or eliminate irrelevant features. One of this preprocessing tasks, feature selection (FS), has turned almost in a must-do, as it can eliminate irrelevant and redundant information, with the added benefits of saving on storage, and allowing for the use of more less complex machine learners, thus improving computational times (see Chap. 2).

Feature selection has been applied to many machine learning and data mining problems, with the aim of selecting a subset of features that minimizes the prediction error of a given classifier. There are different approaches to feature selection, including feature construction, feature ranking, and multivariate feature selection, as well as efficient search methods and feature validity assessment methods [1, 2]. In Chap. 3, we described how better results could be obtained by combining different machine learning methods using an approach called ensemble learning. Ensemble learning has been successfully applied to classification or prediction tasks, but it is also a means for improving feature selection. In this respect, there are two main ways in which ensembles and feature selection are related in the literature. One of them is to use feature selection to provide diversity for posterior ensemble methods, as it is employed in the works of [3, 4]. Another idea, which is the one that will be described in this chapter, is to use ensembles of feature selectors with the goal of improving the stability of the feature selection process [5–10]. This aspect is specially relevant in knowledge discovery, and even more in those cases in which data dimensionality is very high, but the number of samples is not such, as they are more sensible to generalization problems. Thus, several feature selection processes are carried out (either using different training sets, different FS methods, or both), and their results are aggregated to obtain a final subset of features that hopefully will add stability and thus be more transparent in the process of knowledge discovery. The idea is that a more appropriate (stable) feature subset is obtained by combining the multiple feature subsets of the ensemble, as the aggregated result tends to obtain more accurate and stable results, reducing the risk of choosing an unstable subset. If several FS methods are used, the individual selectors in an ensemble are named, by analogy with the base learners, as *base selectors*. If the base selectors are all of the same kind, the ensemble is known as *homogeneous*; otherwise the ensemble is *heterogeneous*. Ensembles can be formed in several ways [11], as we saw before.

As stated above, there are several recent works that have proposed improving feature selection algorithms robustness using multiple feature selection evaluation criteria [12]. The research described in [5] analyzed and compared five measures of diversity for their possible use in ensemble feature selection, and the experiments were carried out over 21 UCI datasets [13], using four different search strategies for ensemble feature selection with simple random subspacing, namely, genetic search, hill-climbing, and ensemble forward and backward sequential selection. In another work, an ensemble consisting on five different base selectors was employed, each selecting a different subset of features that fed five classifiers and their results being combined by simple voting [14]. For improving stability of the FS results beside classification performance, the *Multicriterion Fusion-based Recursive Feature Elimination (MCF-RFE)* algorithm was presented in [15]. Yet another study proposed a feature ranking scheme for *Multilayer Perceptron (MLP)* ensembles [16], used with a stopping criterion based on the *Out-of-Bootstrap (OOB)* estimate [17]; the versatility of this base classifier in removing irrelevant features was demonstrated experimentally using benchmark data. In a scenario as challenging as DNA microarray classification, an ensemble of filters rather than a single filter was used with both synthetic and real data [18].

In another set of works only a specific type of feature selection methods, in this case rankers, were used. In the work described in [19], three filter-based feature ranking techniques with simple combining methods (lowest, highest, and average rank), were employed. Two interesting studies have been carried out by Wang et al. [20, 21]; the first one proposing ensembles of six commonly used filter-based rankers, and the second one studying 17 ensembles of feature ranking techniques with six commonly-used rankers, a signal-to-noise filter technique *(S2N)* [22], and 11 threshold-based rankers. In the second of the Wang et al. studies cited above [21], the ensembles were composed of 2 to 18 individual feature selection methods. Other studies describe different methods for combining individually generated rankings, with the aim of obtaining a final ensemble. The combination of individual rankings covers from simple methods—based on computing the mean, median, minimum, etc.—to more complex methods like *Complete Linear Aggregation* [6, 23] *(CLA)*, *Robust Ensemble Feature Selection (Rob-EFS)* [24], and *SVM-Rank* [6, 12]. Thus, there are two main steps in creating an ensemble for feature selection:

1. Create a set of different feature selectors, each one providing its output. In order to create diversity, there are several methods that can be used, such as using different samples of the training dataset, using different feature selection methods, or a combination of both.
2. Aggregate the results obtained by the single models. There are several measures that can be used in this step, such as majority voting, weighted voting, etc, as was described in depth in Chap. 5. It is important to choose an adequate aggregation method, that is able to preserve the diversity of the individual base models, while maintaining accuracy.

In this chapter we will describe the ideas of the two basic approaches: (i) N selections using the same feature selection algorithm, with each selection using different training data; and (ii) N selections using different feature selection algorithms that use the same training data. The first approach improves computation time by processing data in parallel nodes, whereas the second approach ensures stable and robust feature selection that achieves competitive results irrespective of the scenario.

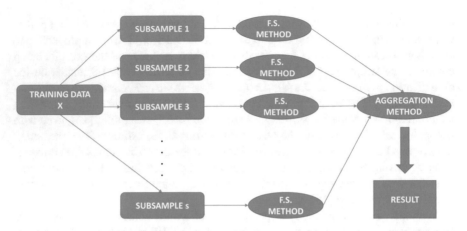

Fig. 4.1 Diagram of an homogeneous distributed ensemble

4.2 Homogeneous Ensembles for Feature Selection

Let us consider a dataset as $\mathbf{X} = \{\mathbf{x_1}, \ldots, \mathbf{x_d}\} \in \mathbb{R}$, with the class label represented as $\mathbf{Y} = \{\mathbf{y_1}, \ldots, \mathbf{y_N}\}$, as it was stated in Chap. 1. Remember that the typical dataset is organized as a matrix of N rows (samples) by d columns (features)– plus an extra column with the class labels. If we draw randomly $\mathbf{s}$ subsamples of size kN, with $0 < k < 1$, where the values of the $\mathbf{s}$ and $\mathbf{k}$ parameters might be varied, we can carry out a feature selection process on each of the $\mathbf{s}$ subsamples. This structure can be seen in Fig. 4.1, in which s models are generated using the same feature selection method, all with different training data.

An important problem of ensemble methods, specially in the present context of Big Data, is the computational time needed in comparison with individual methods. This ensemble structure can be adapted to distributing the training data among a number of nodes. In this way, the training task and feature selection method application can be parallelized, and the final result is obtained thereafter by combining the results obtained in each node using a union method. The pseudo-code of this approach can be seen in Algorithm 4.1.

4.2.1 A Use Case: Homogeneous Ensembles for Feature Selection Using Ranker Methods

We will describe a special case of homogeneous ensemble in which the feature selection method used is a ranker. In this case, as indicated in Algorithm 4.1, the A_n outputs obtained are then combined using a ranking combination method to obtain a

---

Algorithm 4.1: Pseudo-code of an homogeneous distributed ensemble

---

Data: N — number of different nodes

Data: T — threshold of the number of features to be selected (only if a ranker is used)

Result: P — prediction

1 Separate the training data in the N nodes.

2 **for** *each n from 1 to N* **do**

3 $\quad$ Obtaining feature subset (or ranking) A_n using the same feature selection method on the node n

4 A = combining single subsets or rankings A_n with a combination method

5 A_t = Select T top attributes from A (only if a ranker is used)

6 Build classifier/predictor with the selected attributes A_t

7 Obtain prediction P

---

single ranking list. Also, it is necessary to establish a threshold T in order to obtain a practical subset of features, A_t. The code for this homogeneous ensemble is available for downloading.[1]

In order to test the validity of the approach, in the work described in [6], different feature selection methods were employed, three filters and two embedded methods, specifically InfoGain (Information Gain), mRMR(minimum Redundancy Maximum Relevance), ReliefF, SVM-RFE (Recursive Feature Elimination for Support Vector Machines) and FS-P (Feature Selection-Perceptron) (for more details on these methods, please see Chap. 2). Afterwards, a *Support Vector Machine with Radial-Basis-Function (SVM-RBF)* [25] has been employed for checking the adequacy of the proposed ensemble in terms of classification error. Regarding the combination methods used, SVM-Rank, min, median, mean, geoMean, Stuart and RRA (Residual Reduction Algorithm) were employed (see Sect. 5.3 for details on these methods).

Finally, since the feature selection methods used in these approaches are rankers and thus sort all the features, a threshold is necessary to obtain a practical final subset of features. Most works in the literature use several thresholds that retain different percentages of features [26]. Since thresholds are dependent on the particular dataset being studied, several attempts have also been made to derive a general automatic threshold [27, 28]. In this study, five different threshold values were used to delimit data dimensionality, two were automatic thresholds: the $log_2(n)$ threshold, and one proposed in [29] that is based on a data complexity measure, the *Fisher discriminant ratio* [30]. The five different threshold values are the following:

- *Fisher discriminant ratio*: This is defined for a multidimensional problem as:

$$F = \frac{\sum_{i=1, j=1, i \neq j}^{c} p_i p_j (\mu_i - \mu_j)^2}{\sum_{i=1}^{c} p_i \sigma_i^2}, \tag{4.1}$$

---

[1] https://github.com/borjaseijo/dfsre-lib.

where μ_i, σ_i^2, and p_i are the mean, variance, and proportion of the ith class c, respectively. The Fisher discriminant ratio values are calculated individually for each feature of the dataset. In practice, it is preferable to use the *Fisher discriminant ratio* inverse $(1/F)$ to establish the threshold and obtain the final subset, as the smaller value renders the problem more tractable. Therefore, the final formula that calculates the complexity value e of each feature is defined as:

$$e = \alpha \times 1/F + (1 - \alpha) \times \rho \qquad (4.2)$$

where α is a parameter with values in the interval [0, 1] that controls both the relative emphasis on the number of features retained and the weight given to the complexity measure (a value of $\alpha = 0.75$ was empirically established for this work), ρ is the percentage of features retained (ranging from 1 to the total number of features in the dataset) and $1/F$ is the inverse of the *Fisher discriminant ratio*.
- $\log_2(n)$. This threshold, where n is the number of features in a given dataset, following the recommendations in [21, 27].
- *10%*: this threshold selects the 10% of the most relevant features of the final ordered ranking.
- *25%*: this threshold selects the 25% of the most relevant features of the final ordered ranking.
- *50%*: this threshold selects the 50% of the most relevant features of the final ordered ranking.

The ensemble was tested on seven different datasets shown in Table 4.1, as they conform an interesting suite against which to check suitability. The number of samples ranges from 1 484 to 67 557, the number of features oscillates from 8 to 10 000 and the datasets represent both binary and multiclass problems. The experimental procedure, aimed at comparing the individual and homogeneous distributed ensemble approaches using the same rankers, different ranking combination methods, and different training data, was as follows:

Table 4.1 Datasets employed in the experimental study with homogeneous and heterogeneous ensembles

Dataset	Samples	Features	Classes	Download
Yeast	1 484	8	10	UCI repository [13]
Spambase	4 601	57	2	UCI repository [13]
Madelon	2 400	500	2	UCI repository [13]
Connect4	67 557	42	3	UCI repository [13]
Isolet	7 797	617	26	UCI repository [13]
USPS	9 298	256	10	FS repository [31]
Pixraw10P	100	10 000	10	FS repository [31]

1. Each one of the seven datasets (Table 4.1), was partitioned according to a 10-fold cross-validation scheme.
2. The feature selection process was applied as indicated by the ensemble approach (see Algorithm in 4.1).
3. The individual rankings were combined using the different aggregation methods detailed above, so as to obtain the final ranking.
4. A practical subset of features was obtained according to the different thresholds described above.
5. The suitability of the ensemble approach against the individual methods, using a Support Vector Machine (SVM) as classifier to measure the estimated test error was tested. In this study, the *SVM* classifier used a *Gaussian Radial-Basis-Function (RBF)* with values $C = 1$ and *gamma* $= 0.01$ (default values for both parameters in Weka).
6. To the ten different results obtained after 10-fold cross validation, a Kruskal-Wallis test was applied to check if there were significant differences between individual and ensemble strategies for a level of significance $\alpha = 0.05$. Then, a multiple comparison procedure [32] was applied to identify results that were not significantly worse than the best individual result.

Although homogeneous ensembles have interest on their own, in this specific study it was of interest to reduce training time while maintaining classification accuracy. Due to the huge quantity of methods and data sets, average training times, average test errors and standard deviations are shown in the tables that follow.

Table 4.2 shows the average training times in seconds for the five feature selection methods applied to the seven datasets. Individual strategies whose average times were not significantly worse than those of the ensemble strategies using the same feature selection method are labeled with a superscript dagger. As one example, for the *SVM-RFE* method applied to the *Yeast* dataset, the performance of the individual strategy was not significantly worse than that of the ensemble strategy in terms of average training time (the individual and ensemble results are labeled with a superscript dagger), whereas, in contrast, for this method applied to the *Madelon* dataset, the individual strategy was significantly worse than the ensemble strategy (only the ensemble result is labeled with a superscript dagger). Note that the time spent on the ranking combination can be considered negligible. Overall it can be seen how the ensemble strategy considerably improved training times.

Figures 4.2 and 4.3 show the average time gains for the homogeneous distributed ensemble versus the individual approaches. As can be seen, the homogeneous ensembles compared to the individual approaches improved times by a factor of 100 on

Table 4.2 Average training time in seconds taken by the homogeneous distributed ensemble and the individual rankers. The superscript dagger indicates times for individual rankers that were not significantly worse than for the ensembles

Ranker		Yeast	Spambase	Madelon	Connect4	Isolet	USPS	Pixraw10P
InfoGain	Ensem.	0.01† ±0.01	0.01† ±0.01	0.10† ±0.25	0.02† ±0.01	0.06† ±0.01	0.03† ±0.01	0.18† ±0.30
	Indiv.	0.01 ±0.01	0.02† ±0.02	0.02† ±0.01	0.05† ±0.01	0.18† ±0.01	0.05† ±0.01	0.08† ±0.01
mRMR	Ensem.	0.01† ±0.00	8.67† ±2.58	218.35† ±3.91	0.23† ±0.01	25.49† ±0.14	3.06† ±0.21	737.83† ±30.31
	Indiv.	0.01 ±0.01	13.54 ±3.89	510.90 ±25.37	1.38 ±0.07	59.64 ±0.38	25.86 ±3.02	1265.63 ±87.41
ReliefF	Ensem.	0.01† ±0.01	0.01† ±0.00	0.12† ±0.23	0.48† ±0.03	0.20† ±0.01	0.13† ±0.07	0.63† ±0.21
	Indiv.	0.02† ±0.01	0.20 ±0.04	0.69 ±0.04	37.57 ±3.86	8.35 ±0.44	5.88 ±0.90	17.27 ±2.38
SVM-RFE	Ensem.	0.03† ±0.03	0.01† ±0.01	6.51† ±13.51	7.01† ±0.66	37.82† ±9.57	13.5† ±0.11	32.61† ±8.51
	Indiv.	0.05† ±0.03	0.12 ±0.06	1744.28 ±218.17	691.62 ±90.16	2662.18 ±249.78	1082.89 ±121.71	3167.53 ±190.01
FS-P	Ensem.	0.03† ±0.02	0.06† ±0.06	0.54† ±0.04	0.84† ±0.17	18.63† ±0.35	1.59† ±0.31	18.79† ±3.14
	Indiv.	0.30 ±0.09	0.73 ±0.12	4.91 ±0.18	13.60 ±2.43	179.35 ±16.19	17.62 ±0.49	227.91 ±12.84

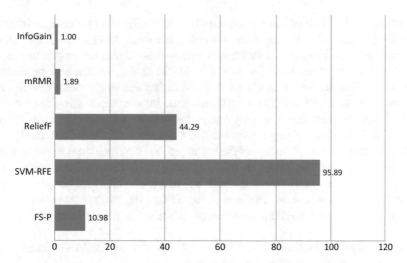

Fig. 4.2 Average speedup for homogeneous distributed ensembles versus individual approaches

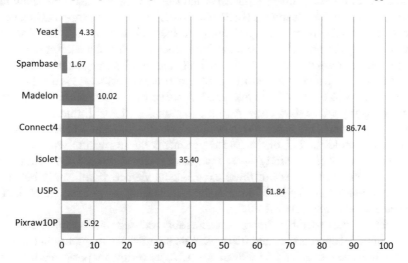

Fig. 4.3 Average speedup for each dataset using the homogeneous distributed ensembles versus individual approaches

average in the best case. The feature selection method whose average training times most improved in the distribution process was the embedded *SVM-RFE*. The fact that the *InfoGain* filter yielded the poorest improvement is not surprising, since it is a univariate and fast method – so even attempts at parallelization produced no improve-ment [33]. Figure 4.3, referring to the datasets, shows that the best time improvement occurred with the *Connect4* dataset; this was because it had the largest number of samples of all the datasets and so was able to take most advantage of the distribu-tion process. For datasets with greater dimensionality and smaller sample sizes, e.g., *Madelon, Isolet. USPS* and *Pixraw10P*, the time improvement with the *SVM-RFE* method was also significant, mainly due to the relatively small number of samples used in each iterative training run by the *SVM* classifier used by this method.

The most important advantage of the homogeneous distributed ensemble approach was clearly the great reduction in training times while classification performance held at reasonable—and sometimes even improved—levels. This outcome reflects the notion of divide-and-conquer since, in some cases, the result obtained by a feature selection method may be more accurate when the focus is on a local region of the data.

Concerning the thresholds, Tables 4.3, 4.4, 4.5, 4.6, and 4.7 show average test errors and standard deviations for the five thresholds, the first two based on auto-matic thresholds (Fisher discriminat ratio and log_2), while the last three show fixed percentages of retained features (10, 25 and 50%). The ensemble approaches whose average test results were not significantly worse than for the individual strategy using the same feature selection method are labeled with a superscript dagger. As can be seen, test error rates for individual and ensemble strategies were compara-ble. Table 4.3 shows that, for 215 of 245 experiments performed with the *Fisher discriminant ratio* threshold, the average test errors for the ensemble strategy were not significantly worse than those for the individual methods; furthermore, in nine of these experiments the ensemble method achieved significantly better results. For the 10% threshold, in 217 of the 245 experiments the test errors returned by the ensemble were not significantly worse than those returned by the individual method (Table 4.5), and a further six of these experiments yielded significantly better test error percentages than the individual method. Increasing the percentage of features retained did not imply better results.

To sum up, the ensemble strategy considerably reduced training times compared to the individual approaches, and there were no significant differences in test errors between the two strategies in 1 096 of the 1 225 experiments performed. In other words, in 89.47% of the cases the performance of the ensemble strategy was not significantly worse than the performance of the individual methods. Note also that, in 30 of the 1 225 experiments, the significant differences were even in favor of the homogeneous distributed ensemble.

Table 4.3 Homogeneous distributed ensemble with a Fisher discriminant ratio threshold: average estimated percentage test errors. The superscript dagger indicates ensemble results that were not significantly different than those for individual rankers

Ranker		Yeast	Spambase	Madelon	Connect4	Isolet	USPS	Pixraw10P
InfoGain	E-SVMRank	58.22 † ±3.95	15.28 † ±2.98	34.75 † ±3.36	30.27 † ±1.73	65.99 ± 6.16	54.50 † ±2.10	82.00 † ±13.77
	E-Min	57.75 † ±4.28	14.85 † ±2.86	33.46 † ±2.84	28.83 † ±2.28	56.88 † ±7.01	56.69 † ±1.76	87.00 † ±12.42
	E-Median	58.22 † ±3.95	16.65 † ±3.55	33.63 † ±3.07	30.76 † ±0.71	62.13 † ±6.58	55.83 † ±2.15	81.00 † ±15.84
	E-Mean	58.22 † ±3.95	15.13 † ±2.51	34.75 † ±3.51	30.76 † ±0.71	65.59 ± 4.91	55.11 † ±2.18	87.00 † ±12.01
	E-GeomMean	58.22 † ±3.95	16.61 † ±2.60	34.00 † ±2.55	30.76 † ±0.71	64.15 † ±5.61	57.98 † ±2.24	84.00 † ±15.09
	E-Stuart	58.22 † ±3.95	13.71 † ±1.99	34.21 † ±2.96	30.25 † ±1.56	65.82 ± 6.44	56.51 † ±1.95	88.00 † ±15.26
	E-RRA	58.22 † ±3.95	14.37 † ±3.63	34.25 † ±3.02	29.80 † ±1.68	65.26 † ±4.41	56.29 † ±1.59	81.00 † ±15.47
	Individual	57.48 † ±4.40	17.93 † ±2.84	34.04 † ±3.48	30.75 † ±0.52	55.05 † ±5.43	56.00 † ±1.64	84.00 † ±14.30
mRMR	E-SVMRank	56.81 † ±3.53	18.71 ± 3.80	52.00 ± 3.34	32.22 † ±2.71	46.90 † ±2.99	18.11 † ±1.98	22.03 † ±18.13
	E-Min	58.22 † ±3.95	15.78 † ±1.54	37.42 † ±3.88	30.77 † ±0.71	47.56 † ±2.38	16.73 † ±0.87	22.04 † ±17.40
	E-Median	56.47 † ±4.12	14.78 † ±1.89	49.42 † ±5.97	32.90 † ±0.89	46.06 † ±2.22	15.10 † ±1.87	23.27 † ±19.75
	E-Mean	57.42 † ±3.78	19.78 ± 3.66	52.75 ± 3.68	32.39 † ±1.86	47.20 † ±2.81	16.07 † ±0.86	23.17 † ±19.50
	E-GeomMean	56.61 † ±3.84	19.17 † ±6.03	49.67 † ±5.56	31.98 † ±0.75	46.19 † ±2.78	15.41 † ±1.23	22.57 † ±18.20
	E-Stuart	57.42 † ±3.78	20.17 ± 2.94	50.58 † ±5.55	33.20 † ±1.11	47.83 ± 3.50	19.45 ± 1.75	21.51 † ±18.48
	E-RRA	57.55 † ±3.79	20.47 ± 2.15	50.38 † ±5.24	33.37 † ±1.01	49.61 ± 2.54	14.43 † ±2.06	23.24 † ±18.34
	Individual	53.44 † ±3.19	13.35 † ±1.69	41.92 † ±1.61	32.27 † ±0.55	43.93 † ±1.08	16.48 † ±1.47	22.00 † ±18.74
ReliefF	E-SVMRank	57.48 † ±3.31	16.41 † ±4.03	33.33 † ±3.07	27.31 ± 1.90	61.05 ± 1.96	25.02 † ±3.80	82.00 † ±19.54
	E-Min	57.96 † ±4.18	16.24 † ±2.01	33.59 † ±2.99	28.21 † ±2.33	61.58 ± 2.08	25.36 † ±3.95	86.00 † ±20.04
	E-Median	57.35 † ±3.95	16.85 † ±3.54	33.21 † ±2.89	27.29 ± 2.01	61.50 ± 1.98	26.61 † ±2.87	87.00 † ±17.51

(continued)

Table 4.3 (continued)

Ranker		Yeast	Spambase	Madelon	Connect4	Isolet	USPS	Pixraw10P
	E-Mean	57.82 † ±3.33	14.41 † ±2.49	33.42 † ±3.05	27.72 † ±2.12	61.50 ± 1.98	25.86 † ±2.22	82.00 † ±18.57
	E-GeomMean	57.82 † ±3.33	14.84 † ±3.09	33.50 † ±2.93	27.72 † ±2.12	61.54 ± 1.93	29.54 ± 1.21	81.00 † ±21.00
	E-Stuart	56.81 † ±5.78	14.74 † ±2.28	33.42 † ±3.05	27.27 ± 1.99	61.24 ± 2.47	26.58 † ±2.81	82.00 † ±17.17
	E-RRA	56.81 † ±5.78	15.28 † ±2.46	33.38 † ±2.91	27.62 ± 2.16	61.24 ± 2.47	25.10 † ±3.18	83.00 † ±22.57
	Individual	57.48 † ±4.40	15.84 † ±2.15	33.33 † ±4.14	30.76 † ±0.72	56.46 † ±1.86	25.75 † ±2.70	86.00 † ±19.55
SVM-RFE	E-SVMRank	49.60 † ±3.72	11.93 † ±1.24	34.00 † ±2.61	33.58 † ±1.76	55.01 † ±2.54	24.38 † ±4.04	26.00 † ±21.34
	E-Min	55.13 † ±4.88	11.69 † ±1.42	33.50 † ±2.46	31.79 † ±3.25	49.99 † ±5.40	23.93 † ±3.46	27.00 † ±22.72
	E-Median	51.42 † ±4.92	11.48 † ±0.73	33.79 † ±2.90	33.25 † ±1.51	53.55 † ±4.28	24.47 † ±3.97	26.00 † ±20.55
	E-Mean	50.07 † ±4.32	12.06 † ±1.21	33.88 † ±2.95	33.17 † ±1.77	54.32 † ±2.93	25.42 † ±4.24	28.00 † ±22.88
	E-GeomMean	50.95 † ±4.86	11.32 † ±0.83	33.33 † ±3.15	33.13 † ±1.22	54.73 † ±2.42	24.75 † ±4.35	26.00 † ±20.43
	E-Stuart	50.07 † ±4.32	11.43 † ±0.86	33.50 † ±3.37	33.85 † ±1.26	55.70 † ±2.53	25.77 † ±3.79	27.00 † ±22.62
	E-RRA	50.07 † ±4.32	11.15 † ±1.05	33.96 † ±3.47	32.14 † ±2.73	54.05 † ±2.57	25.28 † ±3.73	29.00 † ±21.98
	Individual	54.45 † ±4.56	12.82 † ±2.39	35.75 † ±4.36	33.52 † ±0.92	51.53 † ±6.24	16.55 † ±7.92	20.00 † ±24.94
FS-P	E-SVMRank	55.73 † ±3.20	12.26 † ±1.56	34.63 † ±3.60	33.75 † ±1.23	56.89 ± 2.28	12.64 † ±9.62	75.00 † ±19.00
	E-Min	56.75 † ±4.10	11.85 † ±0.70	40.58 † ±7.45	33.56 † ±1.81	55.80 ± 3.32	12.51 † ±9.70	80.00 † ±15.63
	E-Median	57.89 † ±3.93	11.19 † ±0.90	33.38 † ±2.75	33.63 † ±1.86	57.04 † ±1.73	12.51 † ±9.70	82.00 † ±13.17
	E-Mean	56.74 † ±3.49	12.00 † ±1.60	34.88 † ±3.50	34.13 † ±0.90	56.71 ± 1.71	12.51 † ±9.70	80.00 † ±14.91
	E-GeomMean	56.07 † ±3.23	11.39 † ±1.08	33.58 † ±3.20	33.48 † ±1.72	56.77 ± 1.70	12.51 † ±9.70	81.00 † ±12.87
	E-Stuart	55.93 † ±3.38	11.54 † ±1.15	34.00 † ±2.83	33.56 † ±1.94	56.30 ± 1.31	12.51 † ±9.70	82.00 † ±13.98
	E-RRA	55.93 † ±3.38	12.65 † ±2.05	34.96 † ±2.90	34.02 † ±1.05	56.51 ± 2.47	12.51 † ±9.70	78.00 † ±15.49
	Individual	54.18 † ±3.28	12.19 † ±1.74	34.71 † ±3.32	33.43 † ±2.31	60.25 † ±2.23	8.74 † ±3.95	78.00 † ±15.49

Table 4.4 Homogeneous distributed ensemble with a $\log_2(n)$ threshold: average estimated percentage test errors. The superscript dagger indicates ensemble results that were not significantly different than those for individual rankers

Ranker		Yeast	Spambase	Madelon	Connect4	Isolet	USPS	Pixraw10P
InfoGain	E-SVMRank	57.62† ±5.05	13.28† ±0.66	34.58 ±2.14	30.77† ±0.53	67.62 ± 1.59	57.88† ±2.09	80.00† ±22.62
	E-Min	57.29† ±5.03	12.93† ±1.63	33.54† ±2.07	30.85† ±0.52	61.38† ±5.53	62.43 ± 2.91	85.00† ±20.68
	E-Median	57.62† ±5.05	13.11† ±1.06	33.67† ±2.02	30.75† ±0.52	64.10† ±4.25	53.76† ±2.92	86.00† ±22.82
	E-Mean	57.62† ±5.05	12.80† ±1.38	37.50 ±6.87	30.75† ±0.52	67.14 ±1.47	58.48† ±1.31	84.00† ±20.12
	E-GeomMean	57.62† ±5.05	13.17† ±0.62	34.08† ±2.26	30.75† ±0.52	66.94† ±1.97	56.79† ±2.94	85.00† ±21.10
	E-Stuart	57.62† ±5.05	12.93† ±1.40	34.21† ±2.10	30.78† ±0.56	67.30 ±1.68	53.58† ±2.91	84.00† ±20.18
	E-RRA	57.62† ±5.05	12.87† ±1.68	34.46† ±1.60	30.81† ±0.55	66.13† ±2.41	54.67† ±1.97	82.00† ±20.38
	Individual	58.76† ±3.92	13.41† ±1.87	33.83† ±4.70	30.77† ±0.52	58.23† ±6.83	56.00† ±1.64	82.00† ±21.50
mRMR	E-SVMRank	55.06† ±4.67	21.84† ±1.55	50.79 ± 2.49	32.98† ±0.65	47.52† ±1.75	25.79† ±1.95	24.00† ±19.66
	E-Min	57.62† ±5.05	29.32† ±2.31	36.92† ±4.27	30.76† ±0.53	47.02† ±3.56	26.56† ±1.34	26.00† ±17.14
	E-Median	55.47† ±5.26	26.97† ±2.53	48.46† ±5.09	32.91† ±0.45	45.86† ±2.58	26.83† ±1.58	25.00† ±19.02
	E-Mean	56.89† ±5.45	21.21† ±1.68	50.38 ± 2.90	33.08† ±0.41	47.75 ±1.66	25.18† ±1.22	24.00† ±19.01
	E-GeomMean	56.34† ±4.72	24.89† ±2.09	49.71† ±4.32	32.23† ±0.65	46.57† ±3.03	29.55 ± 1.75	27.00† ±17.52
	E-Stuart	56.89† ±5.45	22.78† ±1.75	49.75† ±4.46	33.08† ±0.41	48.38† ±1.97	23.59† ±1.25	26.00† ±18.27
	E-RRA	56.28† ±5.69	22.41† ±1.94	50.54 ± 5.50	33.09† ±0.41	49.96† ±1.78	24.03† ±1.50	25.00† ±16.30
	Individual	53.91† ±3.52	22.82† ±1.99	42.17† ±2.77	32.22† ±0.35	43.49† ±1.92	24.79† ±1.14	24.00† ±17.76
ReliefF	E-SVMRank	56.42† ±6.38	16.54† ±2.85	33.46† ±2.13	30.09† ±0.69	61.23† ±1.87	29.21† ±1.01	87.00† ±18.91
	E-Min	56.08† ±6.29	14.80† ±2.64	33.50† ±2.11	30.46† ±0.77	61.45† ±2.74	36.12 ± 1.33	85.00† ±21.65
	E-Median	57.62† ±5.05	17.08† ±2.70	33.42† ±1.98	30.03† ±0.73	61.78† ±1.92	32.11† ±1.16	87.00† ±18.91

(continued)

Table 4.4 (continued)

Ranker		Yeast	Spambase	Madelon	Connect4	Isolet	USPS	Pixraw10P
	E-Mean	57.02 † ±6.38	14.80 † ±3.21	33.54 † ±2.13	30.06 † ±0.73	61.78 † ±1.92	32.73 † ±1.79	87.00 † ±18.91
	E-GeomMean	57.02 † ±6.38	14.93 † ±3.59	33.58 † ±2.21	30.06 † ±0.73	61.87 † ±1.81	29.51 † ±1.31	87.00 † ±18.91
	E-Stuart	56.75 † ±6.45	15.08 † ±2.79	33.54 † ±2.13	30.15 † ±0.63	61.58 † ±2.30	38.27 ± 1.52	86.00 † ±18.60
	E-RRA	57.35 † ±5.17	14.37 † ±1.70	33.50 † ±2.05	30.02 † ±0.69	61.58 † ±2.30	30.87 † ±1.16	84.00 † ±21.30
	Individual	58.76 † ±3.92	21.28 † ±3.10	33.21 † ±4.24	30.74 † ±0.49	59.47 † ±1.71	31.10 † ±1.78	86.00 † ±19.55
SVM-RFE	E-SVMRank	49.60 † ±3.56	11.67 † ±1.88	33.83 † ±1.43	33.95 † ±0.83	53.60 † ±2.47	33.46 ± 2.66	50.00 † ±23.49
	E-Min	54.05 † ±3.90	12.04 † ±1.12	33.71 † ±2.10	34.14 † ±0.51	51.11 † ±5.41	26.13 † ±3.29	36.00 † ±23.73
	E-Median	50.27 † ±3.51	11.39 † ±1.30	33.96 † ±1.43	33.53 † ±0.94	50.44 † ±2.14	26.13 † ±3.29	34.00 † ±23.95
	E-Mean	49.60 † ±3.56	11.80 † ±2.34	33.83 † ±1.51	33.87 † ±0.76	53.25 † ±2.19	29.12 † ±4.10	35.00 † ±24.66
	E-GeomMean	50.27 † ±3.51	11.37 † ±1.30	33.92 † ±1.79	34.09 † ±0.55	51.44 † ±2.24	27.21 † ±3.36	34.00 † ±23.95
	E-Stuart	49.60 † ±3.56	11.17 † ±1.83	33.46 † ±2.04	34.16 † ±0.49	53.94 ± 2.32	28.75 † ±4.32	32.00 † ±23.19
	E-RRA	49.60 † ±3.56	11.06 † ±1.48	33.71 † ±2.01	34.18 † ±0.48	53.23 † ±1.68	29.68 † ±2.86	37.00 † ±26.77
	Individual	55.86 † ±4.44	12.54 † ±2.58	33.50 † ±4.35	33.90 † ±0.44	48.12 † ±5.98	26.00 † ±3.33	32.00 † ±25.73
FS-P	E-SVMRank	55.47 † ±4.96	11.84 † ±1.90	33.88 † ±1.64	34.13 † ±0.49	56.25 ± 2.44	53.81 † ±4.67	75.00 † ±19.00
	E-Min	56.01 † ±5.71	11.48 † ±1.88	43.17 † ±8.19	34.12 † ±0.47	55.43 ± 2.81	53.81 † ±4.67	87.00 † ±13.37
	E-Median	58.09 † ±5.11	11.63 † ±1.81	34.00 † ±1.78	34.15 † ±0.50	56.80 ± 2.02	53.81 † ±4.67	86.00 † ±10.75
	E-Mean	56.82 † ±6.62	12.74 † ±2.38	33.83 † ±1.86	34.13 † ±0.50	55.86 ± 2.27	53.81 † ±4.67	87.00 † ±11.60
	E-GeomMean	55.81 † ±5.93	11.35 † ±1.39	34.17 † ±1.75	34.14 † ±0.52	56.54 ± 2.24	53.81 † ±4.67	84.00 † ±11.74
	E-Stuart	56.88 † ±4.70	11.50 † ±1.85	33.58 † ±1.68	34.14 † ±0.49	55.84 ± 2.01	53.81 † ±4.67	84.00 † ±11.74
	E-RRA	56.88 † ±4.70	12.00 † ±2.00	33.58 † ±1.71	34.09 † ±0.54	56.71 ± 2.23	53.81 † ±4.67	82.00 † ±12.29
	Individual	54.38 † ±3.11	12.00 † ±1.90	33.50 † ±4.42	34.16 † ±0.41	61.86 † ±1.33	57.18 † ±7.10	81.00 † ±8.76

Table 4.5 Homogeneous distributed ensemble with a 10% threshold: average estimated percentage test errors. The superscript dagger indicates ensemble results that were not significantly different than those for individual rankers

Ranker		Yeast	Spambase	Madelon	Connect4	Isolet	USPS	Pixraw10P
InfoGain	E-SVMRank	59.16 † ±4.31	13.06 † ±2.00	34.38 † ±3.66	30.80 † ±0.79	49.17 † ±1.64	47.88 † ±2.09	83.00 † ±20.14
	E-Min	55.12 † ±2.57	13.11 † ±1.97	34.42 † ±3.28	30.73 † ±0.81	48.34 † ±1.91	50.43 ± 2.91	79.00 † ±23.39
	E-Median	58.62 † ±4.46	13.15 † ±1.65	34.38 † ±3.23	30.76 † ±0.76	49.31 † ±2.18	43.76 † ±2.92	81.00 † ±23.73
	E-Mean	60.65 † ±3.78	13.06 † ±1.96	34.58 † ±3.71	30.76 † ±0.76	48.90 † ±1.68	50.48 ± 1.31	83.00 † ±22.71
	E-GeomMean	59.50 † ±3.88	13.06 † ±1.87	34.29 † ±3.40	30.76 † ±0.76	48.92 † ±1.67	46.79 † ±2.94	83.00 † ±23.03
	E-Stuart	61.73 ± 3.30	13.00 † ±1.99	34.42 † ±3.58	30.72 † ±0.73	48.76 † ±1.53	43.58 † ±2.91	84.00 † ±22.97
	E-RRA	61.79 ± 3.39	12.80 † ±1.71	34.38 † ±3.45	30.71 † ±0.72	48.70 † ±1.19	44.67 † ±1.97	82.00 † ±21.56
	Individual	55.13 † ±4.99	13.39 † ±1.24	33.62 † ±3.50	30.76 † ±0.54	48.62 † ±2.30	45.38 † ±1.68	80.00 † ±21.08
mRMR	E-SVMRank	55.12 † ±2.57	22.02 † ±2.90	51.25 † ±3.09	32.97 † ±0.82	50.01 † ±1.87	14.29 † ±1.45	28.00 † ±21.52
	E-Min	55.12 † ±2.57	29.99 ± 2.14	38.59 † ±6.73	30.82 † ±0.87	49.44 † ±1.43	13.77 † ±0.84	29.00 † ±19.76
	E-Median	55.12 † ±2.57	27.52 ± 1.91	49.75 † ±3.53	32.92 † ±0.71	49.99 † ±2.36	12.26 † ±1.08	30.00 † ±21.44
	E-Mean	55.12 † ±2.57	21.28 † ±2.64	50.54 † ±3.14	33.08 † ±0.81	49.94 † ±1.98	14.80 † ±0.72	27.00 † ±19.73
	E-GeomMean	55.12 † ±2.57	24.69 † ±1.50	47.83 † ±3.12	32.14 † ±0.71	50.12 † ±1.83	11.13 † ±1.25	25.00 † ±21.78
	E-Stuart	55.12 † ±2.57	23.06 † ±2.33	47.75 † ±3.72	33.08 † ±0.81	50.07 † ±2.62	12.75 † ±0.75	25.00 † ±21.78
	E-RRA	55.12 † ±2.57	22.52 † ±3.08	47.33 † ±3.71	33.08 † ±0.81	49.52 † ±2.79	12.52 † ±1.00	26.00 † ±19.58
	Individual	55.13 † ±4.99	22.78 † ±2.24	46.42 † ±3.46	32.29 † ±0.49	47.15 † ±1.71	12.94 † ±0.96	26.00 † ±20.66
ReliefF	E-SVMRank	55.12 † ±2.57	16.32 † ±3.79	33.67 † ±3.87	30.10 † ±0.64	57.23 † ±2.68	17.50 † ±1.91	69.00 † ±20.34
	E-Min	55.12 † ±2.57	14.65 ± 2.91	33.46 † ±3.99	30.30 † ±0.77	56.51 † ±2.65	19.69 † ±1.28	72.00 † ±19.04
	E-Median	55.12 † ±2.57	17.56 † ±4.39	33.71 † ±3.88	30.09 † ±0.59	57.40 † ±2.71	18.83 † ±1.75	74.00 † ±19.64

(continued)

Table 4.5 (continued)

Ranker		Yeast	Spambase	Madelon	Connect4	Isolet	USPS	Pixraw10P
	E-Mean	55.12 † ±2.57	15.30 ± 3.87	34.25 † ±3.88	30.08 † ±0.66	57.29 † ±2.64	18.83 † ±1.75	75.00 † ±22.17
	E-GeomMean	55.12 † ±2.57	15.61 ± 4.09	33.83 † ±4.42	30.08 † ±0.66	57.09 † ±2.65	21.98 ± 1.38	70.00 † ±21.11
	E-Stuart	55.12 † ±2.57	15.67 ± 3.68	34.08 † ±3.40	30.04 † ±0.65	56.89 † ±2.77	19.51 † ±1.56	72.00 † ±19.04
	E-RRA	55.12 † ±2.57	14.52 ± 3.03	34.04 † ±3.85	30.05 † ±0.71	56.83 † ±2.78	19.29 † ±1.07	72.00 † ±19.04
	Individual	55.13 † ±4.99	20.08 † ±3.14	33.17 † ±3.13	30.70 † ±0.50	58.38 † ±2.23	18.79 † ±1.36	71.00 † ±20.79
SVM-RFE	E-SVMRank	51.22 † ±3.36	12.06 † ±2.07	34.46 † ±3.37	33.89 † ±0.97	64.74 ± 3.77	10.23 † ±1.58	20.00 † ±19.05
	E-Min	55.47 † ±5.81	12.12 † ±1.96	34.00 † ±3.89	34.16 † ±0.72	52.73 † ±3.23	11.56 ± 1.42	18.00 † ±19.05
	E-Median	51.42 † ±3.33	11.67 † ±1.99	34.67 † ±3.28	33.54 † ±0.95	62.25 ± 3.04	10.06 † ±1.27	20.00 † ±20.81
	E-Mean	51.29 † ±3.37	12.19 † ±2.07	34.58 † ±3.32	33.84 † ±1.05	65.56 ± 3.24	11.55 ± 0.98	21.00 † ±20.81
	E-GeomMean	51.29 † ±3.37	11.54 † ±1.98	34.63 † ±3.46	33.99 † ±0.84	61.79 ± 3.75	10.10 † ±0.93	19.00 † ±19.86
	E-Stuart	51.42 † ±3.33	11.80 † ±2.17	34.46 † ±3.30	34.17 † ±0.69	63.61 ± 2.26	10.87 ± 0.94	20.00 † ±20.62
	E-RRA	50.88 † ±3.65	11.56 † ±1.85	34.83 † ±3.15	34.17 † ±0.68	64.23 ± 3.72	11.34 ± 0.80	21.00 † ±20.81
	Individual	54.31 † ±6.50	12.50 † ±1.41	31.71 † ±2.56	33.92 † ±0.59	51.58 † ±3.33	8.32 † ±0.70	15.00 † ±18.41
FS-P	E-SVMRank	54.59 † ±4.54	11.43 † ±1.44	34.13 † ±3.61	34.13 † ±0.69	72.44 ± 3.27	15.58 † ±3.51	84.00 † ±17.13
	E-Min	54.99 † ±4.13	11.78 † ±2.78	35.63 † ±5.15	34.13 † ±0.70	71.69 † ±3.64	16.03 † ±3.41	60.00 † ±20.00
	E-Median	55.80 † ±5.14	11.63 † ±2.00	34.04 † ±4.30	34.12 † ±0.67	72.78 ± 3.37	16.03 † ±3.41	60.00 † ±20.00
	E-Mean	55.06 † ±3.52	12.50 † ±2.20	34.79 † ±3.06	34.15 † ±0.72	72.50 ± 3.30	16.03 † ±3.41	60.00 † ±20.00
	E-GeomMean	55.12 † ±3.54	11.34 † ±2.09	35.21 † ±4.29	34.12 † ±0.71	72.67 ± 3.22	16.03 † ±3.41	61.00 † ±21.32
	E-Stuart	55.06 † ±3.52	11.43 † ±1.74	34.33 † ±3.31	34.14 † ±0.71	72.36 ± 3.10	16.03 † ±3.41	60.00 † ±23.09
	E-RRA	55.12 † ±2.57	11.95 † ±1.70	33.96 † ±4.19	34.09 † ±0.69	71.71 † ±2.88	16.03 † ±3.41	60.00 † ±23.09
	Individual	54.66 † ±4.33	12.17 † ±1.52	33.96 † ±2.94	34.18 † ±0.60	64.95 † ±4.53	16.80 † ±2.22	63.00 † ±24.97

Table 4.6 Homogeneous distributed ensemble with a 25% threshold: average estimated percentage test errors. The superscript dagger indicates ensemble results that were not significantly different than those for individual rankers

Ranker		Yeast	Spambase	Madelon	Connect4	Isolet	USPS	Pixraw10P
InfoGain	E-SVMRank	59.16 † ±4.31	18.06 † ±1.32	36.71 † ±3.16	26.43 † ±0.66	43.18 † ±2.47	21.61 † ±1.52	73.00 † ±25.34
	E-Min	55.12 † ±2.57	18.06 † ±1.48	36.75 † ±3.58	26.41 † ±0.67	42.72 † ±2.62	21.91 † ±2.20	67.00 † ±26.71
	E-Median	58.62 † ±4.46	18.21 † ±1.30	37.29 † ±3.36	26.41 † ±0.67	43.38 † ±2.58	19.75 † ±2.42	72.00 † ±24.54
	E-Mean	60.65 ± 3.78	18.06 † ±1.32	36.96 † ±3.70	26.41 † ±0.67	43.16 † ±2.49	18.44 † ±1.44	72.00 † ±24.54
	E-GeomMean	59.50 † ±3.88	18.06 † ±1.32	37.21 † ±4.22	26.41 † ±0.67	43.21 † ±2.51	19.63 † ±1.23	72.00 † ±24.54
	E-Stuart	61.73 ± 3.30	18.19 † ±1.74	36.33 † ±3.70	26.41 † ±0.67	43.54 † ±2.54	19.63 † ±1.23	68.00 † ±26.68
	E-RRA	61.79 ± 3.39	18.37 † ±1.72	36.96 † ±3.49	26.41 † ±0.67	43.32 † ±2.76	20.37 † ±1.63	69.00 † ±24.42
	Individual	53.17 † ±6.89	17.67 † ±1.90	36.42 † ±3.59	26.34 † ±0.72	43.68 † ±1.94	19.48 † ±2.44	69.00 † ±26.44
mRMR	E-SVMRank	55.12 † ±2.57	16.54 † ±1.50	52.17 † ±2.63	31.50 † ±0.85	48.92 † ±3.01	9.53 ± 0.61	35.00 † ±20.95
	E-Min	55.12 † ±2.57	18.34 ± 1.62	40.17 † ±7.64	29.07 ± 1.05	49.34 † ±3.02	9.14 † ±1.48	35.00 † ±20.95
	E-Median	55.12 † ±2.57	16.61 † ±1.61	51.25 † ±2.63	31.58 † ±1.14	48.97 † ±3.08	9.71 ± 1.00	33.00 † ±22.91
	E-Mean	55.12 † ±2.57	16.74 † ±1.73	52.83 † ±2.19	31.75 † ±0.87	48.51 † ±3.04	9.67 ± 0.95	31.00 † ±21.85
	E-GeomMean	55.12 † ±2.57	16.76 † ±1.68	48.00 † ±2.53	30.46 † ±1.36	48.61 † ±3.09	8.00 † ±1.20	31.00 † ±19.67
	E-Stuart	55.12 † ±2.57	16.80 † ±1.73	48.92 † ±2.79	31.90 † ±1.11	48.07 † ±3.24	9.89 ± 1.21	35.00 † ±20.95
	E-RRA	55.12 † ±2.57	16.56 † ±1.04	48.96 † ±2.92	33.53 † ±0.66	47.99 † ±3.17	7.59 † ±1.23	30.00 † ±19.23
	Individual	53.17 † ±6.89	15.65 † ±1.95	49.17 † ±1.66	32.01 † ±0.72	46.72 † ±2.71	7.24 † ±1.00	32.00 † ±20.98
ReliefF	E-SVMRank	55.12 † ±2.57	17.11 † ±2.26	35.67 † ±3.71	25.56 † ±0.69	56.91 † ±2.43	27.88 † ±0.90	53.00 † ±20.20
	E-Min	55.12 † ±2.57	15.06 ± 2.81	36.54 † ±3.09	25.56 † ±0.69	54.84 † ±2.29	25.59 † ±0.86	51.00 † ±21.80
	E-Median	55.12 † ±2.57	17.54 † ±2.17	36.21 † ±3.71	25.56 † ±0.69	55.71 † ±2.63	27.63 † ±0.65	52.00 † ±21.66

(continued)

Table 4.6 (continued)

Ranker		Yeast	Spambase	Madelon	Connect4	Isolet	USPS	Pixraw10P
	E-Mean	55.12 † ±2.57	16.84 † ±1.97	35.79 † ±4.32	25.56 † ±0.69	57.41 † ±2.33	27.07 † ±0.16	52.00 † ±21.66
	E-GeomMean	55.12 † ±2.57	16.82 † ±1.85	36.83 † ±2.90	25.56 † ±0.69	56.87 † ±2.34	28.89 † ±0.26	55.00 † ±21.15
	E-Stuart	55.12 † ±2.57	16.80 † ±1.97	36.58 † ±3.26	25.57 † ±0.68	56.38 † ±2.38	27.59 † ±0.34	52.00 † ±21.66
	E-RRA	55.12 † ±2.57	16.93 † ±1.98	35.92 † ±3.03	25.57 † ±0.68	55.78 † ±2.70	28.20 ± 0.78	55.00 ± 21.15
	Individual	53.17 † ±6.89	19.34 † ±3.69	36.58 † ±3.47	25.62 † ±0.53	54.08 † ±3.64	25.62 † ±0.53	50.00 ± 24.50
SVM-RFE	E-SVMRank	51.22 † ±3.36	11.56 † ±4.21	38.04 † ±2.41	33.11 † ±0.91	57.84 ± 3.71	7.51 † ±0.88	23.00 † ±21.53
	E-Min	55.47 † ±5.81	12.54 † ±4.33	35.42 ± 3.89	33.99 † ±0.74	45.94 † ±3.45	7.68 † ±1.23	23.00 † ±22.33
	E-Median	51.42 † ±3.33	11.71 † ±4.63	36.38 ± 3.02	33.11 † ±1.06	54.71 ± 3.52	7.13 † ±1.43	25.00 † ±21.16
	E-Mean	51.29 † ±3.37	10.89 ± 3.96	37.25 † ±2.84	32.65 † ±1.27	58.59 ± 3.04	9.99 ± 1.46	25.00 † ±21.16
	E-GeomMean	51.29 † ±3.37	12.02 † ±4.62	36.54 ± 2.93	33.24 † ±1.09	54.98 ± 2.45	9.24 ± 0.81	23.00 † ±22.33
	E-Stuart	51.42 † ±3.33	12.54 † ±4.51	36.58 ± 2.98	33.60 † ±1.07	58.65 ± 3.23	9.67 ± 1.21	23.00 † ±22.33
	E-RRA	50.88 † ±3.65	12.54 † ±4.54	36.25 ± 2.56	33.56 † ±1.25	59.93 ± 3.33	9.52 ± 1.78	25.00 † ±22.46
	Individual	53.56 † ±6.45	18.04 † ±3.99	41.83 † ±2.82	33.16 † ±1.18	42.20 † ±1.77	5.56 † ±0.69	22.00 † ±22.01
FS-P	E-SVMRank	54.59 † ±4.53	10.13 † ±3.05	36.58 † ±2.65	34.07 † ±0.72	75.94 ± 4.45	6.63 † ±1.16	69.00 † ±14.49
	E-Min	54.99 † ±4.13	12.41 † ±5.15	36.29 † ±3.74	34.17 † ±0.69	69.44 † ±4.90	6.63 † ±1.16	56.00 † ±21.19
	E-Median	55.80 † ±5.14	11.67 † ±3.91	37.25 † ±2.17	34.17 † ±0.69	74.45 ± 4.03	6.63 † ±1.16	55.00 † ±21.21
	E-Mean	55.06 † ±3.52	10.15 † ±3.13	36.96 † ±3.01	34.16 † ±0.70	75.43 ± 4.39	6.63 † ±1.16	56.00 † ±21.19
	E-GeomMean	55.12 † ±3.54	10.80 † ±3.83	36.58 † ±2.41	34.17 † ±0.69	75.00 ± 4.17	6.63 † ±1.16	56.00 † ±21.19
	E-Stuart	55.06 † ±3.52	10.00 † ±3.07	36.08 † ±2.06	34.07 † ±0.73	74.86 ± 4.08	6.63 † ±1.16	55.00 † ±20.68
	E-RRA	55.12 † ±2.57	9.67 † ±3.00	35.75 ± 2.84	33.82 † ±0.64	75.07 ± 3.76	6.63 † ±1.16	55.00 † ±20.68
	Individual	53.03 † ±4.51	14.45 † ±5.53	39.17 † ±1.94	34.03 † ±0.54	61.45 † ±4.34	7.27 ± 1.15	59.00 † ±20.25

Table 4.7 Homogeneous distributed ensemble with a 50% threshold: average estimated percentage test errors. The superscript dagger indicates ensemble results that were not significantly different than those for individual rankers

Ranker		Yeast	Spambase	Madelon	Connect4	Isolet	USPS	Pixraw10P
InfoGain	E-SVMRank	59.16 † ±4.31	16.89 † ±1.20	38.50 † ±3.72	24.78 † ±0.55	37.98 † ±3.34	14.93 ± 1.13	57.00 † ±28.55
	E-Min	55.12 † ±2.57	16.76 † ±1.35	39.29 † ±4.11	24.97 † ±0.69	36.73 † ±2.67	12.92 † ±1.02	59.00 † ±26.13
	E-Median	58.62 † ±4.46	16.87 † ±1.14	38.50 † ±3.56	24.76 † ±0.70	38.46 † ±2.93	11.37 † ±1.68	57.00 † ±26.27
	E-Mean	60.65 † ±3.78	16.89 † ±1.21	39.04 † ±3.94	24.79 † ±0.52	38.05 † ±3.40	12.33 † ±0.54	60.00 † ±27.63
	E-GeomMean	59.50 † ±3.88	16.93 † ±1.20	39.38 † ±3.53	24.80 † ±0.45	38.01 † ±3.12	10.43 † ±1.35	59.00 † ±28.12
	E-Stuart	61.73 ± 3.30	16.93 † ±1.17	38.13 † ±3.98	25.09 † ±0.89	38.34 † ±2.52	13.62 † ±1.12	60.00 † ±27.63
	E-RRA	61.79 ± 3.39	17.02 † ±1.16	38.04 † ±3.40	25.22 † ±0.94	38.17 † ±2.40	13.51 † ±0.82	60.00 † ±27.63
	Individual	54.05 † ±4.31	16.95 † ±1.28	38.54 † ±3.49	24.90 † ±0.79	37.98 † ±2.36	11.79 † ±1.10	58.00 † ±28.21
mRMR	E-SVMRank	55.12 † ±2.57	14.13 † ±1.17	39.96 † ±2.96	29.75 † ±0.80	43.50 † ±3.18	7.33 ± 0.55	34.00 † ±23.89
	E-Min	55.12 † ±2.57	15.37 † ±1.21	40.08 † ±2.97	28.27 ± 1.25	42.02 † ±3.69	6.27 † ±1.41	37.00 † ±24.67
	E-Median	55.12 † ±2.57	14.08 † ±1.05	39.67 † ±3.02	30.55 † ±0.87	42.81 † ±3.73	5.27 † ±1.25	35.00 † ±25.02
	E-Mean	55.12 † ±2.57	14.11 † ±1.11	40.17 † ±3.33	29.82 † ±0.76	42.66 † ±3.49	5.79 † ±1.53	37.00 † ±24.67
	E-GeomMean	55.12 † ±2.57	14.06 † ±1.06	39.83 † ±2.81	28.99 ± 0.66	42.62 † ±3.79	5.46 † ±1.38	37.00 † ±24.67
	E-Stuart	55.12 † ±2.57	14.24 † ±1.02	40.21 † ±4.11	30.20 † ±0.79	43.71 † ±2.91	7.84 ± 0.77	33.00 † ±25.37
	E-RRA	55.12 † ±2.57	16.13 ± 2.07	40.25 † ±3.50	31.05 † ±1.02	43.91 † ±3.26	6.32 † ±1.41	35.00 † ±25.02
	Individual	54.05 † ±4.31	13.28 † ±1.79	39.21 † ±3.48	31.01 † ±0.51	46.75 † ±3.36	5.36 † ±0.69	36.00 † ±24.59
ReliefF	E-SVMRank	55.12 † ±2.57	17.84 † ±0.85	38.92 † ±1.86	23.39 † ±0.66	47.02 † ±3.95	7.10 ± 1.04	37.00 † ±23.07
	E-Min	55.12 † ±2.57	17.34 † ±1.23	39.09 † ±2.54	23.39 † ±0.59	45.52 † ±4.28	6.16 † ±1.14	35.00 † ±22.36
	E-Median	55.12 † ±2.57	17.76 † ±0.91	40.33 † ±3.73	23.23 † ±0.64	46.11 † ±3.79	7.74 ± 1.06	36.00 † ±23.15

(continued)

Table 4.7 (continued)

Ranker		Yeast	Spambase	Madelon	Connect4	Isolet	USPS	Pixraw10P
	E-Mean	55.12 † ±2.57	17.82 † ±0.96	38.79 † ±1.74	23.22 † ±0.64	47.04 † ±3.92	5.00 † ±0.84	38.00 † ±22.75
	E-GeomMean	55.12 † ±2.57	17.61 † ±0.90	39.58 † ±2.32	23.22 † ±0.64	46.36 † ±4.08	6.38 † ±0.57	35.00 † ±22.36
	E-Stuart	55.12 † ±2.57	17.91 ± 0.80	38.71 † ±2.16	23.48 † ±0.67	46.30 † ±4.25	6.27 † ±0.85	38.00 † ±23.07
	E-RRA	55.12 † ±2.57	18.08 ± 0.69	38.71 † ±2.77	23.39 † ±0.76	45.91 † ±4.12	6.38 † ±0.94	35.00 † ±22.36
	Individual	54.05 ±4.31	16.17 † ±1.15	39.92 † ±2.89	23.51 † ±0.43	50.33 † ±3.45	5.56 † ±0.69	37.00 † ±23.12
SVM-RFE	E-SVMRank	51.22 † ±3.36	18.02 † ±1.11	39.46 † ±3.51	31.34 † ±2.03	48.71 † ±4.14	5.62 † ±0.95	46.00 † ±22.47
	E-Min	55.47 † ±5.81	17.67 † ±1.16	39.00 † ±3.05	31.37 † ±1.86	39.98 † ±1.62	7.70 ± 1.16	33.00 † ±21.47
	E-Median	51.42 † ±3.33	18.06 † ±1.11	39.75 † ±3.25	31.79 † ±2.45	48.08 † ± 4.18	5.35 † ±1.27	35.00 † ±23.55
	E-Mean	51.29 † ±3.37	17.97 † ±1.11	39.21 † ±3.01	31.37 † ±2.02	48.92 ± 3.63	7.12 ± 0.85	34.00 † ±22.91
	E-GeomMean	51.29 † ±3.37	17.82 † ±1.29	39.71 † ±2.70	30.51 † ±1.64	44.57 † ±3.29	5.70 † ±1.16	34.00 † ±22.91
	E-Stuart	51.42 † ±3.33	17.82 † ±1.00	39.38 † ±2.95	31.49 † ±1.60	48.36 ± 4.29	6.35 † ±0.91	33.00 † ±22.27
	E-RRA	50.88 † ±3.65	18.04 † ±1.13	39.17 † ±2.92	31.85 † ±1.48	50.31 † ±4.73	7.83 ± 1.34	36.00 † ±23.17
	Individual	52.57 † ±6.65	18.11 † ±1.66	42.25 † ±3.23	31.80 † ±2.35	37.70 † ±3.12	4.98 † ±0.66	33.00 † ±23.12
FS-P	E-SVMRank	54.59 † ±4.54	17.15 † ±3.23	39.63 † ±2.33	34.07 ± 0.64	65.99 ± 6.19	5.19 † ±0.85	61.00 † ±20.79
	E-Min	54.99 † ±4.13	17.26 † ±2.65	38.75 † ±2.28	34.17 ± 0.69	60.06 † ±6.32	5.19 † ±0.85	49.00 † ±23.78
	E-Median	55.80 † ±5.14	18.08 † ±1.06	39.21 † ±3.28	34.06 ± 0.67	65.24 ± 6.32	5.19 † ±0.85	48.00 † ±22.01
	E-Mean	55.06 † ±3.52	18.06 † ±1.15	39.17 † ±2.43	34.08 ± 0.65	66.14 ± 6.26	5.19 † ±0.85	48.00 † ±22.01
	E-GeomMean	55.12 † ±3.54	18.24 † ±1.12	38.54 † ±2.90	34.12 ± 0.69	64.77 ± 6.65	5.19 † ±0.85	49.00 † ±23.78
	E-Stuart	55.06 † ±3.52	18.34 † ±1.07	39.00 † ±2.07	34.04 ± 0.64	65.51 ± 6.61	5.19 † ±0.85	48.00 † ±22.01
	E-RRA	55.12 † ±2.57	18.15 † ±1.23	39.71 † ±3.61	33.82 † ±1.12	65.33 ± 6.24	5.19 † ±0.85	48.00 † ±22.01
	Individual	54.80 † ±5.87	16.06 † ±4.18	40.67 † ±3.84	32.55 † ±0.96	48.39 † ±4.95	5.22 † ±0.63	53.00 † ±23.12

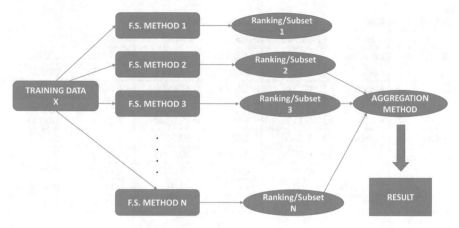

Fig. 4.4 A structure for an heterogeneous ensemble

4.3 Heterogeneous Ensembles for Feature Selection

In this case, the diversity is introduced by using N models employing different feature selection methods, but the same training data (Fig. 4.4). This approach takes account of the strengths and weaknesses of the individual methods, with the objective also of devising a more robust result. The several different methods are trained using the same training data, and the output is then combined using an aggregation method. The pseudo-code of this approach is given in Algorithm 4.2.

Algorithm 4.2: Pseudo-code of an heterogeneous ensemble

Data: N — number of ranker methods
Data: T — threshold of the number of features to be selected (only for ranker methods)

Result: P — classification prediction

1 **for** *each n from 1 to N* **do**
2   Obtaining ranking or subset of features A_n using feature selection method n

3 A = combining rankings or feature subsets A_n with an aggregator
4 A_t = Select T top attributes from A, only if rankers are used
5 Build classifier SVM-RBF with the selected attributes A_t
6 Obtain prediction P

Table 4.8 ρ value of Spearman's rank correlation coefficient

Dataset	Ranker	InfoGain	mRMR	ReliefF	SVM-RFE	FS-P
Spambase	InfoGain	1.0000	0.2011	0.0714	−0.2040	−0.1736
	mRMR	0.2011	1.0000	−0.0811	0.1313	0.0838
	ReliefF	0.0714	−0.0811	1.0000	−0.0672	0.0380
	SVM-RFE	−0.2040	0.1313	−0.0672	1.0000	0.0565
	FS-P	−0.1736	0.0838	0.0380	0.0565	1.0000
Isolet	InfoGain	1.0000	0.0971	−0.0677	−0.0320	−0.0521
	mRMR	0.0971	1.0000	0.0295	0.0534	0.0062
	ReliefF	−0.0677	0.0295	1.0000	0.0115	−0.0291
	SVM-RFE	−0.0320	0.0534	0.0115	1.0000	0.0331
	FS-P	−0.0521	0.0062	−0.0291	0.0331	1.0000

4.3.1   A Use Case: Heterogeneous Ensemble for Feature Selection Using Ranker Methods

In this case, a heterogeneous ensemble was devised using the five different FS methods named for the previous homogeneous use case, that is, InfoGain, mRMR, ReliefF, SVM-RFE and FS-P. Remember that all of them, the two embedded and three filters, are rankers, and thus in Algorithm 4.2 we should apply the thresholds and aggregators that are applicable for rankers (see previous Sect. 4.2.1). The code for this heterogeneous ensemble is available for downloading.[2]

This particular set of rankers was selected because: (i) they are based on different metrics and thus it is expected to ensure great diversity in the final ensemble; and (ii) they are widely used by feature selection researchers. In Table 4.8 a small diversity study using just two of the datasets, (*Spambase* and *Isolet*) is shown in Table 4.1. Final rankings obtained by the five rankers were compared using Spearman's rank correlation coefficient [34]. The ρ value in the range $[-1, 1]$ reflects the relationship between rankings, with 1 indicating that the compared rankings were equal.

It can be seen that most of the ρ values are far from 1, indicating great differences between the paired rankings (obviously, when the same ranker method rankings were compared, the ρ value was 1, as can be seen in the table diagonals). This small experiment demonstrated that the set of feature selection rankers chosen for this study ensured enough diversity in their behaviors. For more details see [6].

In the case of the heterogeneous ensemble, the main aim is to maintain or improve classification performance while freeing the user from having to decide on the most appropriate feature selection method for any given situation. Average test errors and standard deviations for the five thresholds employed are shown in Tables 4.9, 4.10, and 4.11. The best results obtained are for the *Fisher discriminant ratio* and 10% of retained features, as in the previous use case for the homogeneous design (see

---

[2]https://github.com/borjaseijo/fsre-lib.

Table 4.9 Heterogeneous centralized ensemble with a *Fisher discriminant ratio* threshold: average estimated percentage test errors. The superscript dagger indicates results that were not significantly different from the best result

Ranker	Yeast	Spambase	Madelon	Connect4	Isolet	USPS	Pixraw10P
E-SVMRank	54.78 [†] ±3.27	11.26 [†] ±1.52	34.75 [†] ±5.16	30.06 [†] ±2.02	52.06 [†] ±4.82	31.21 ± 5.21	77.00 ± 20.58
E-Min	53.44 [†] ±3.19	18.93 ± 3.05	34.29 [†] ±3.33	31.36 [†] ±1.11	51.21 [†] ±1.79	18.04 [†] ±6.13	60.00 [†] ±17.64
E-Median	56.00 [†] ±2.95	11.67 [†] ±1.57	33.21 [†] ±4.33	29.98 [†] ±1.81	54.57 ± 4.15	37.70 ± 4.90	85.00 ± 19.00
E-Mean	54.78 [†] ±3.27	12.82 [†] ±3.16	34.46 [†] ±4.50	29.84 [†] ±2.20	49.72 [†] ±3.51	30.47 ± 4.70	44.00 [†] ±24.13
E-GeomMean	53.44 [†] ±3.19	11.28 [†] ±1.54	33.67 [†] ±3.93	30.90 [†] ±0.94	52.25 [†] ±3.67	36.01 ± 3.45	60.00 [†] ±23.09
E-Stuart	54.78 [†] ±3.27	11.28 [†] ±1.54	33.63 [†] ±3.94	30.30 [†] ±1.94	51.94 [†] ±3.40	35.74 ± 4.08	50.00 [†] ±29.06
E-RRA	56.47 [†] ±5.04	14.13 [†] ±3.37	33.17 [†] ±3.95	29.69 [†] ±2.00	50.30 [†] ±5.44	28.53 [†] ±4.06	59.00 [†] ±21.32
InfoGain	57.48 [†] ±4.40	17.93 ± 2.84	34.04 [†] ±3.48	30.75 [†] ±0.52	55.05 ± 5.43	56.00 ± 1.64	84.00 ± 14.30
mRMR	53.44 [†] ±3.19	13.35 [†] ±1.69	41.92 ± 1.61	32.27 [†] ±0.55	43.93 [†] ±1.08	16.48 [†] ±1.47	22.00 [†] ±18.74
ReliefF	57.48 [†] ±4.40	15.84 [†] ±2.15	33.33 [†] ±4.14	30.76 [†] ±0.72	56.46 ± 1.86	25.75 [†] ±2.70	86.00 ± 19.55
SVM-RFE	54.45 [†] ±4.56	12.82 [†] ±2.39	35.75 [†] ±4.36	33.52 ± 0.92	51.53 [†] ±6.24	16.55 [†] ±7.92	20.00 [†] ±24.94
FS-P	54.18 [†] ±3.28	12.19 [†] ±1.74	34.71 [†] ±3.32	33.43 ± 2.31	60.25 ± 2.23	8.74 [†] ±3.95	78.00 ± 15.49

Sect. 4.2.1). The algorithms whose average test error results were not significantly worse than the best result are labeled with a superscript dagger, as in the previous sections. For more detailed results, please consult [6].

As it can be seen in the previous tables, the experimental results demonstrated the suitability of the proposed ensemble, since they matched or improved on the results achieved by the individual feature selection methods. The ensemble errors were not significantly different from the lowest average error for the individual methods in 40 of the 49 experiments performed with the *Fisher discriminant ratio* threshold (Table 4.9). The ensemble method obtained favorable results in 42 of the 49 experiments performed with a 10% threshold (Table 4.11). Figure 4.5 shows the number of cases for which the results obtained by the individual and the heterogeneous ensemble approaches were not significantly different than the best result (in other words, the number of times that results were comparable with the best result). As can be observed, the *E-RRA* ensemble approach obtained results that were not significantly different from the best result in all 35 experiments, compared to 28 out of 35 exper-

Table 4.10 Heterogeneous centralized ensemble with a $\log_2(n)$ threshold: average estimated percentage test errors. The superscript dagger indicates results that were not significantly different from the best result

Ranker	Yeast	Spambase	Madelon	Connect4	Isolet	USPS	Pixraw10P
E-SVMRank	54.58 [†] ±3.33	12.22 [†] ±2.98	33.71 [†] ±4.82	31.30 [†] ±0.71	50.58 [†] ±3.94	31.21 [†] ±2.48	77.00 ± 20.58
E-Min	53.91 [†] ±3.52	20.04 ± 1.78	33.46 [†] ±4.40	32.28 ± 0.60	46.80 [†] ±3.14	33.82 [†] ±4.90	65.00 [†] ±20.68
E-Median	57.62 [†] ±3.20	11.76 [†] ±1.09	33.46 [†] ±4.45	31.48 [†] ±0.44	55.36 ± 4.76	37.70 ± 4.70	85.00 ± 19.00
E-Mean	54.24 [†] ±3.22	14.76 [†] ±4.52	33.50 [†] ±4.30	31.40 [†] ±0.68	51.26 [†] ±4.62	30.47 [†] ±3.45	44.00 [†] ±20.66
E-GeomMean	53.91 [†] ±3.52	12.19 [†] ±2.92	33.33 [†] ±4.37	31.25 [†] ±0.73	51.73 [†] ±3.26	36.01 ± 6.08	56.00 [†] ±26.75
E-Stuart	54.65 [†] ±4.16	11.30 [†] ±1.37	33.33 [†] ±4.37	31.22 [†] ±0.78	50.57 [†] ±3.99	35.74 ± 4.06	54.00 [†] ±27.97
E-RRA	56.20 [†] ±3.94	13.67 [†] ±3.15	33.50 [†] ±4.44	30.97 [†] ±0.78	50.97 [†] ±4.16	28.53 [†] ±0.00	70.00 [†] ±19.44
InfoGain	58.76 [†] ±3.92	13.41 [†] ±1.87	33.83 [†] ±4.70	30.77 [†] ±0.52	58.23 ± 6.83	56.00 ± 1.64	82.00 ± 21.50
mRMR	53.91 [†] ±3.52	22.82 ± 1.99	42.17 ± 2.77	32.22 ± 0.35	43.49 [†] ±1.92	24.79 [†] ±1.14	24.00 [†] ±17.76
ReliefF	58.76 [†] ±3.92	21.28 [†] ±3.10	33.21 [†] ±4.24	30.74 [†] ±0.49	59.47 ± 1.71	31.10 [†] ±1.78	86.00 ± 19.55
SVM-RFE	55.86 [†] ±4.44	12.54 [†] ±2.58	33.50 [†] ±4.35	33.90 ± 0.44	48.12 [†] ±5.98	26.00 [†] ±3.33	32.00 [†] ±25.73
FS-P	54.38 [†] ±3.11	12.00 [†] ±1.90	33.50 [†] ±4.42	34.16 ± 0.41	61.86 ± 1.33	57.18 ± 7.10	81.00 ± 8.76

iments for the best performing individual feature selection method (*SVM-RFE*). Five of the remaining six ensemble methods (*E-SVMRank*, *E-Median*, *E-Mean*, *E-GeomMean* and *E-Stuart*) matched or (mostly) improved on the results obtained by *SVM-RFE*, obtaining results that were not significantly different in 28-33 of the 35 experiments.

Focusing on the behavior of the individual feature selection rankers (the bottom five rows in each of the two tables), it can be observed that not one of the five individual methods significantly outperformed the ensemble approaches for any dataset or threshold combination. Therefore, although an individual method might well perform better than an ensemble method in a given scenario, overall it would appear that the ensemble approach is the most consistent and reliable approach to a feature selection process. Overall, an ensemble approach would seem to be the most reliable approach to feature selection, although in some specific cases, an individual method (not always the same one) might well perform better than the ensemble.

Table 4.11 Heterogeneous centralized ensemble with a 10% threshold: average estimated percentage test errors. The superscript dagger indicates results that were not significantly different from the best result

Ranker	Yeast	Spambase	Madelon	Connect4	Isolet	USPS	Pixraw10P
E-SVMRank	54.18 † ±2.72	11.26 † ±1.52	36.29 † ±3.12	31.40 † ±0.87	51.49 † ±1.93	14.09 † ±0.98	84.00 ± 14.30
E-Min	50.27 † ±3.97	19.63 ± 2.46	35.21 † ±2.79	32.09 ± 0.64	58.48 ± 2.27	15.98 ± 2.52	47.00 † ±27.10
E-Median	54.18 † ±2.72	11.67 † ±1.57	36.63 † ±3.86	31.51 † ±0.66	52.79 † ±2.57	13.18 † ±0.70	50.00 † ±21.60
E-Mean	54.18 † ±2.72	13.02 † ±3.70	36.21 † ±3.61	31.36 † ±0.75	51.22 † ±1.81	14.12 † ±1.19	30.00 † ±18.26
E-GeomMean	54.18 † ±2.72	11.28 † ±1.54	37.25 † ±3.82	31.20 † ±0.78	54.89 ± 2.10	14.36 ± 1.60	48.00 † ±25.58
E-Stuart	54.18 † ±2.72	11.28 † ±1.54	37.04 † ±4.26	31.16 † ±0.70	51.65 † ±1.15	13.65 † ±1.56	43.00 † ±24.06
E-RRA	54.18 † ±2.72	14.50 † ±4.12	36.54 † ±3.50	31.01 † ±0.44	50.29 † ±1.36	13.56 † ±1.13	47.00 † ±24.52
InfoGain	55.13 † ±4.99	13.39 † ±1.24	33.62 † ±3.50	30.76 † ±0.54	48.62 † ±2.30	45.38 ± 1.68	80.00 ± 21.08
mRMR	55.13 † ±4.99	22.78 ± 2.24	46.42 ± 3.46	32.29 ± 0.49	47.15 † ±1.71	12.94 † ±0.96	26.00 † ±20.66
ReliefF	55.13 † ±4.99	20.08 ± 3.14	33.17 † ±3.13	30.70 † ±0.50	58.38 ± 2.23	18.79 ± 1.36	71.00 ± 20.79
SVM-RFE	54.31 † ±6.50	12.50 † ±1.41	31.71 † ±2.56	33.92 ± 0.59	51.58 † ±3.33	8.32 † ±0.70	15.00 † ±18.41
FS-P	54.66 † ±4.33	12.17 † ±1.52	33.96 † ±2.94	34.18 ± 0.60	64.95 ± 4.53	16.80 ± 2.22	63.00 ± 24.97

4.4 A Comparison on the Result of Both Use Cases: Homogeneous Versus Heterogeneous Ensemble for Feature Selection Using Ranker Methods

Finally, and in order to give some recommendations in the use of both approaches, Fig. 4.6 shows a graphical comparison between the homogeneous distributed and the heterogeneous centralized ensembles. The figure compares the two best heterogeneous centralized ensembles in terms of average test error (*E-Mean* and *E-RRA*), and the single best homogeneous distributed ensemble in terms of average training times (*SVM-RFE*). Note that the combination methods used by the homogeneous distributed ensemble were also the *Mean* and the *RRA* functions.

As can be seen in Fig. 4.6, the homogeneous distributed ensemble obtained significantly better results than the heterogeneous centralized ensemble for the *Yeast* dataset – the smallest of all the datasets in terms of both size and dimension. As

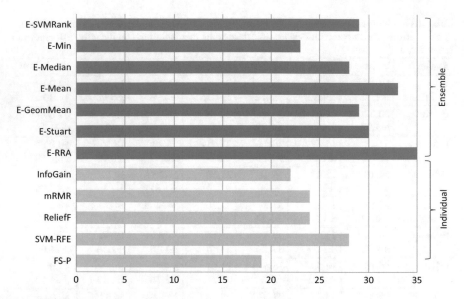

Fig. 4.5 Number of cases when the results obtained by the individual and the heterogeneous centralized ensemble approaches were comparable with the best result

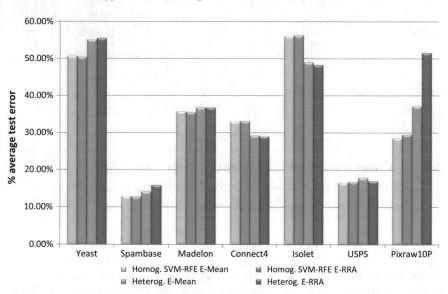

Fig. 4.6 Comparison of average estimated percentage test errors for homogeneous distributed and heterogeneous centralized ensembles

dataset size and dimension increased, however, the difference in test error between the two ensemble types diminished to the point of reversing the situation. Thus, the best result for the heterogeneous centralized ensemble was obtained for the *Connect4* and *Isolet* datasets; this is explained by the fact that the homogeneous approach can improve computation time—at the cost of a minimal reduction in accuracy—by distributing the sample over different nodes. Finally, when the *RRA* combination method was used, a very high average test error was returned for the *Pixraw10P* dataset, a result that was improved on slightly when the *Mean* combination method was used. As it is well-known, microarray datasets are especially susceptible to great accuracy variations depending on the combination method used.

To sum up, we propose applying the following rules-of-thumb:

- If the dataset is very large and a reduction in training time is crucial, the homogeneous distributed ensemble is the best option, since it considerably reduces training time while ensuring reasonable classification accuracy.
- If the dataset is reasonably small, or the user is uncertain as to which of the available algorithms to choose, the heterogeneous centralized ensemble is probably the best option, since it does not require the user to decide between feature selection methods, and may even, in some cases, improve classification accuracy.

4.5   Summary

In this chapter we have described the fundamentals that have been used for adapting the basic ideas of ensemble learning to the feature selection process, aiming at taking advantage of the combination of different individual feature selection methods. The homogeneous ensemble uses the same FS method, but different training sets, and thus computational time can be greatly reduced by parallelizing the training task. This advantage beside aiding diversity to the process can help to make it possible using feature selection methods in Big Data scenarios, some times restricted due to the low scalability of traditional FS algorithms. In the case of the heterogeneous approach, which consists of using different feature selection methods for the same training data, the idea tries to take advantage of the strengths and overcome the weaknesses of the individual methods. The latter approach has the added benefit of freeing the user from the task of deciding which method best suits a particular scenario. Both approaches shown competitive results without deteriorating classification accuracy. As a suggestion, the homogeneous distributed ensemble is particularly suitable for large datasets, while the heterogeneous centralized ensemble has the advantage of freeing the user from decision making regarding the best possible feature selection method for a given problem.

References

1. Guyon, I., Elisseeff, A.: An introduction to variable and feature selection. J. Mach. Learn. Res. **3**, 1157–1182 (2003)
2. Hall, M.A., Holmes, G.: Benchmarking attribute selection techniques for discrete class data mining. IEEE Trans. Knowl. Data Eng. **15**(6), 1437–1447 (2003)
3. Cunningham, P., Carney, J.: Diversity versus quality in classification ensembles based on feature selection. In: Lpez de Mntaras, R., Plaza, E. (eds.) European Conference on Machine Learning (ECML). LNAI, vol. 1810, pp. 109–116 (2000)
4. Opitz, D.W.: Feature selection for ensembles. In: Proceedings of the 16th National Conference on Artificial Intelligence, pp. 379–384. AAAI Press (1999)
5. Tsymbal, A., Pechenizkiy, M., Cunningham, P.: Diversity in search strategies for ensemble feature selection. Inf. fusion **6**(1), 1566–2535 (2005)
6. Seijo-Pardo, B., Porto-Díaz, I., Bolón-Canedo, V., Alonso-Betanzos, A.: Ensemble feature selection: homogeneous and heterogeneous approaches. Knowl. Based Syst. **118**, 124–139 (2017). https://doi.org/10.1016/j.knosys.2016.11.017
7. Saeys, Y., Abeel, T., Van der Peer, Y.: Robust feature selection using ensemble feature selection techniques. In: Daelemans, W., et al. (eds.) European Conference on Machine Learning (ECML PKDD). LNAI 5212, pp. 313–325 (2008)
8. Das, A.K., Das, S., Ghosh, A.: Ensemble feature selection using bi-objective genetic algorithm. Knowl. Based Syst. **123**, 116–127 (2017)
9. Tuv, E., Borisov, A., Runger, G., Torkkola, K.: Feature selection with ensembles, artificial variables and redundancy elimination. J. Mach. Learn. Res. **10**, 1241–1366 (2009)
10. Rogers, J.D., Gunn, S.R.: Ensemble algorithms for feature selection. In: Winkler, J., Niranjan, M., Lawrence, N. (eds.) Deterministic and Statistical Methods in Machine Learning. Lecture Notes in Computer Science, vol. 3635. Springer, Berlin (2005)
11. Bramer, M.: Principles of Data Mining. Springer, Berlin (2007)
12. Seijo-Pardo, B., Bolón-Canedo, V., Alonso-Betanzos, A.: Testing different ensemble configurations for feature selection. Neural Process. Lett. **46**, 857–880 (2017)
13. Lichman, M.: UCI machine learning repository. University of California, Irvine, School of Information and Computer Sciences. http://archive.ics.uci.edu/ml. Last Accessed (2018)
14. Bolón-Canedo, V., Sánchez-Maroño, N., Alonso-Betanzos, A.: An ensemble of filters and classifiers for microarray data classification. Pattern Recognit. **45**(1), 531–539 (2012)
15. Yang, F., Mao, K.Z.: Robust feature selection for microarray data based on multicriterion fusion. IEEE/ACM Trans. Comput. Biol. Bioinform. (TCBB) **8**(4), 1080–1092 (2011)
16. Windeatt, T., Duangsoithong, R., Smith, R.: Embedded feature ranking for ensemble MLP classifiers. IEEE Trans. Neural Netw. **22**(6), 988–994 (2011)
17. Windeatt, T., Prior, M.: Stopping criteria for ensemble-based feature selection. Multiple Classifier Systems, pp. 271–281. Springer, Berlin (2007)
18. Bolón-Canedo, V., Sánchez-Maroño, N., Alonso-Betanzos, A.: Data classification using an ensemble of filters. Neurocomputing **135**, 13–20 (2014)
19. Olsson, J., Oard, D.W.: Combining feature selectors for text classification. In: Proceedings of the 15th ACM International Conference on Information and Knowledge Management, pp. 798–799. ACM (2006)
20. Wang, H., Khoshgoftaar, T.M., Gao, K.: Ensemble feature selection technique for software quality classification. In: International Conference on Software Engineering, SEKE 2010, pp. 215–220
21. Wang, H., Khoshgoftaar, T.M., Napolitano, A.: A comparative study of ensemble feature selection techniques for software defect prediction. In: 2010 Ninth International Conference on Machine Learning and Applications (ICMLA), pp. 135–140
22. Yang, C.H., Huang, C.C., Wu, K.C., Chang, H.Y.: A novel GA-Taguchi-based feature selection method. In: Intelligent Data Engineering and Automated Learning–IDEAL 2008, pp. 112–119

23. Abeel, T., Helleputte, T., Van de Peer, Y., Dupont, P., Saeys, Y.: Robust biomarker identification for cancer diagnosis with ensemble feature selection methods. Bioinformatics **26**(3), 392–398 (2010)
24. Ben Brahim, A., Limam, M.: Robust ensemble feature selection for high dimensional data sets. In: 2013 International Conference on High Performance Computing and Simulation (HPCS), pp. 151–157
25. Park, J., Sandberg, I.W.: Universal approximation using radial-basis-function networks. Neural Comput. **3**(2), 246–257 (1991)
26. Bolón-Canedo, V., Sánchez-Maroño, N., Alonso-Betanzos, A.: Recent advances and emerging challenges of feature selection in the context of big data. Knowl. Based Syst. **86**, 33–45 (2015)
27. Khoshgoftaar, T.M., Golawala, M., Van Hulse, J.: An empirical study of learning from imbalanced data using random forest. In: 2007 19th IEEE International Conference on Tools with Artificial Intelligence, ICTAI 2007, Vol. 2, pp. 310–317. IEEE (2007)
28. Mejía-Lavalle, M., Sucar, E., Arroyo, G.: Feature selection with a perceptron neural net. In: Proceedings of the International Workshop on Feature Selection for Data Mining, pp. 131–135 (2006)
29. Morán-Fernández, L., Bolón-Canedo, V., Alonso-Betanzos, A.: Centralized vs. distributed feature selection methods based on data complexity measures. Knowl. Based Syst. **117**, 27–45 (2017)
30. Basu, M., Ho, T.K.: Data Complexity in Pattern Recognition. Springer Science & Business Media, Berlin (2006)
31. Li, J., Cheng, K., Wang, S., Morstatter, F., Trevino, R., Tang, J., Liu, H.: Feature selection: a data perspective (2016). arXiv:1601.07996
32. Tukey, J.W.: Comparing individual means in the analysis of variance. Biometrics **5**, 99–114 (1949)
33. Eiras-Franco, C., Bolón-Canedo, V., Ramos, S., Gónzález-Domínguez, J., Alonso-Betanzos, A., Touriño, J.: Multithreaded and spark parallelization of feature selection filters. J. Comput. Sci. **17**, 609–619 (2016). https://doi.org/10.1016/j.jocs.2016.07.002
34. Lyerly, S.B.: The average Spearman rank correlation coefficient. Psychometrika **17**(4), 421–428 (1952)

Chapter 5
Combination of Outputs

Abstract Ensemble learning is based on the divided-and-conquer principles but, after dividing, we would need to combine the partial results in some way to reach a final decision. Therefore, a crucial point when designing an ensemble method is to choose an appropriate method for combining the different weak outputs. There are several methods in the literature to solve this issue, and they are grouped according to whether the outputs are classification predictions, subsets of features or rankings of features. In this chapter we will describe methods falling in all these categories, so that the interesting readers can make an informed choice according to their needs trying to design the best ensemble possible.

A crucial point when using an ensemble of learning methods is to combine the partial outputs in order to obtain a *final* output. In the concrete case of an ensemble of feature selectors, there are two possible situations: combining the different features selected by the different selectors or, if a classifier is applied after feature selection, combining the label predictions of the classifiers. Figure 5.1a shows an example of ensemble in which the feature selection process is followed by a classifier, so in this case it is necessary to combine the classifier predictions (see Sect. 5.1). In other cases, as shown in Fig. 5.1b, the outputs of the feature selection methods have to be combined before obtaining a final subset of features which are fed to the classifier. As mentioned in Chap. 2, a feature selection method can return a subset of relevant features or an ordered ranking of all the features. Depending on this, the combination of the outputs will be different, as will be discussed in Sects. 5.2 and 5.3.

5.1 Combination of Label Predictions

Combining the outputs of the individual classifiers in a ensemble is a recurrent issue in the field of ensemble learning, since it is necessary when designing an ensemble of classifiers and has been broadly studied (see, for example, Kuncheva's book [1]). Before diving into the different methods to combine classifier predictions, it is necessary to distinguish between two types of classifier outputs:

© Springer International Publishing AG, part of Springer Nature 201883
V. Bolón-Canedo and A. Alonso-Betanzos, *Recent Advances in Ensembles for Feature Selection*, Intelligent Systems Reference Library 147, https://doi.org/10.1007/978-3-319-90080-3_5

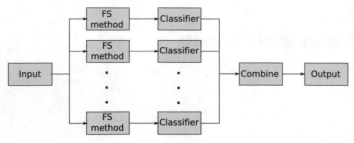

(a) Combination of classifier predictions

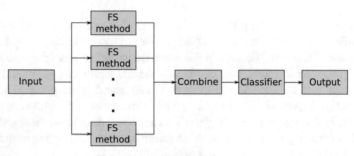

(b) Combination of selected features

Fig. 5.1 Two different examples of feature selection ensembles

- Class labels. In this case, each classifier produces a class label for each data point, without information about the certainty of the guessed labels. All the existing classifiers are able of producing a class label, so all classifiers belong to this category.
- Degree of certainty. Apart from the class label, there are some classifiers that are able to provide a degree of certainty of their prediction (e.g. probability). In this case, we can use the prediction labels in a more informed way and it is possible to combine the outputs giving more importance to those which have a higher degree of certainty.

Depending on the type of classifier outputs, different methods for combining the outputs can be used. When having classifiers that only return the class labels, the most popular technique is *majority vote*.

5.1.1   Majority Vote

The idea behind majority vote is simple and well-known: it consists of establishing the final output as the option that has been predicted by the majority of the classifiers. However, it has some limitations, as for example how to deal with ties, which are

Fig. 5.2 Examples of different scenarios in majority vote

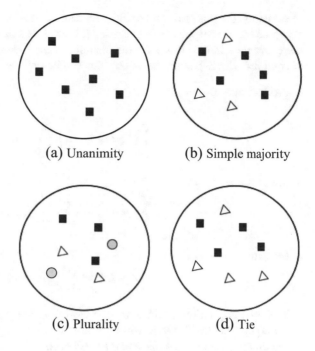

(a) Unanimity (b) Simple majority

(c) Plurality (d) Tie

usually resolved arbitrarily. Figure 5.2 illustrates several situations in which the final output might be the same, but as it is easy to see, the degree of agreement between classifiers is far from being the same.

Notice that in the three first situations (unanimity, simple majority and plurality), the outcome would be the same —the black square—, but the confidence on this outcome is not the same. In Fig. 5.2d we can see an example of a tie, and in this case the final outcome depends on how ties are implemented in our system.

As can be seen, majority vote —although widely employed— has an important number of limitations and problems. Some of them can be solved when combining outputs of classifiers that include a degree of certainty. In the next subsection we present the decision rules that can be used, according to the survey presented by Peteiro-Barral and Guijarro-Berdiñas [2].

5.1.2 Decision Rules

As mentioned before, some classifiers are able to provide a class prediction and also a degree of certainty of this prediction. Consider a classification problem in which instance x has to be assigned to one of the C possible classes of the problem $c_1, c_2, \ldots, c_C$. Let us also assume that we have N classifiers which will lead to N outputs $y_i, i = 1, \ldots, N$ to make the decision. When the classifiers provide a degree of certainty, the posterior probability can be estimated as $P(c_j|x) = y_i$, where y_i is computed as the response of a classifier i. Now, let us denote $y_{ij}(x)$ as the output of the classifier i in the class j for the instance x and assuming that the outputs y_i are normalized. Some popular decision rules can be defined as follows:

- Product rule, $x \rightarrow c_j$ if

$$\prod_{i=1}^{N} y_{ij}(x) = \max_{k=1}^{C} \prod_{i=1}^{N} y_{ik}(x)$$

- Sum rule, $x \rightarrow c_j$ if

$$\sum_{i=1}^{N} y_{ij}(x) = \max_{k=1}^{C} \sum_{i=1}^{N} y_{ik}(x)$$

- Max rule, $x \rightarrow c_j$ if

$$\max_{i=1}^{N} y_{ij}(x) = \max_{k=1}^{C} \max_{i=1}^{N} y_{ik}(x)$$

This rule approximates the sum rule assuming that the output classes are a priori equiprobable. The sum will be dominated by the prediction which lends the maximum support for a particular hypothesis.

- Min rule, $x \rightarrow c_j$ if

$$\min_{i=1}^{N} y_{ij}(x) = \max_{k=1}^{C} \min_{i=1}^{N} y_{ik}(x)$$

This rule approximates the product rule assuming that the output classes are a priori equiprobable. The product will be dominated by the prediction which have the minimum support for a particular hypothesis.

- Median rule, $x \rightarrow c_j$ if

$$\frac{1}{N} \sum_{i=1}^{N} y_{ij}(x) = \max_{k=1}^{C} \frac{1}{N} \sum_{i=1}^{N} y_{ik}(x)$$

5.2   Combination of Subsets of Features

An alternative to the previous approach is to combine the partial results of the ensemble in the feature selection part, and then classifying only once (if classification is the final goal of our system). In this case, we need to integrate the partial feature selections obtained by the different weak selectors.

As mentioned in Chap. 2, feature selection methods can be grouped whether their output is a subset of features or a ranking of all the features. This section will be focused on the former case, and the next section on the latter.

5.2.1 *Intersection and Union*

The most straightforward technique to combine subsets of selected features is to compute the intersection and the union of them. The intersection consists of selecting only those features which are selected by *all* the weak selectors. Notice that the rationale behind this method is very logical, since one can expect that if a feature is selected by all selectors, it has a high predictive power. However, this can lead to very restrictive sets of features, even leading to the empty set of features, and in practice it does not tend to produce good results [3, 4].

On the contrary, the union consists of combining those features selected by *any* weak selector. In this case, the final set of features contains all the features that had been considered important by any selector, but it may lead to select even the whole set of features. This approach tends to produce better results than the intersection [3], but at the expense of a lower reduction in the number of the features.

Imagine a simple toy example in which we have a dataset with five features, $\{a, b, c, d, e\}$, where all of them are relevant but redundant with each other, so it is enough to select any of them. Suppose now that we have five weak selectors and each one selects one of the features (e.g. the first selector chooses feature a, the second selector chooses feature b, and so on). All the weak selectors would have done a perfect job, since any single feature is relevant and it is enough and optimal to select only one of them. But let us show what happens if we combine the results using the intersection or the union:

- If we compute the intersection, the final set of selected features would be the empty subset, which cannot solve the problem.
- If we compute the union, the final set of selected features would be the full set of features, $\{a, b, c, d, e\}$, so it would be correct to solve the problem, but suboptimal because we would be using more features than we actually need.

Of course this is an extremely naïve example, but it helps to illustrate the short-comings of these two combination methods. In the next subsections we can see other more sophisticated techniques, but notice that they come at the expense of higher computational costs.

5.2.2 *Using Classification Accuracy*

Based on the philosophy of a wrapper for feature selection, we can use the classification accuracy (which is usually the ultimate measure of quality of the selected features) to combine the partial subsets of features returned by the weak selectors. A simple approach can be to include partial selections into the final selection only if they contribute to improve classification performance, as presented in [5]. In the method they propose, the first selection S_1 is arbitrarily taken to calculate the classification accuracy, which will be the *baseline*, and the features in S_1 will always become part of the final selection S. For the remaining selections, the features in $S_i, i = 2 \ldots n$ will become part of the final selection S if they improve the baseline accuracy, as can be seen in more detail in Algorithm 5.1. The authors expect that combining the features in this manner can help reduce redundancy, since a redundant feature will not improve the accuracy and hence will not be added to the final selection.

Algorithm 5.1: Pseudocode to use classification accuracy to join subsets of features

Data: $D_{(m \times s)}$ = training dataset with m samples and s features

       n = number of weak selectors
Result: S = final subset of features
1 **for** $i = 1$ *to n* **do**
2 | S_i = subset of features selected by the ith weak selector
  end
3 $S = S_1$
4 *baseline* = accuracy obtained by classifying subset $D_{(m \times |S_1|)}$ with classifier C
5 **for** $i = 2$ *to n* **do**
6 | $S_{aux} = S \cup S_i$
7 | *accuracy* = accuracy obtained by classifying subset $D_{(m \times |S_{aux}|)}$ with classifier C
  | **if** *accuracy > baseline* **then**
8 | | $S = S_{aux}$
9 | | *baseline = accuracy*
  | **end**
  end

This approach has several problems, the most important one is that it depends on how good is the first selection, since it always is part of the final selection. To solve this, it might be more efficient to retain only those features that are selected more times by the weak selectors, making use of classification accuracy and trying to reduce to the extent possible the percentage of selected features, as proposed in [6].

The idea is simple: each time that a weak selector performs feature selection, those features not selected receive a vote. Eventually, the features that have received a number of votes above a certain threshold are removed. Determining the threshold of votes is not trivial, since it depends on the dataset. Therefore, they have proposed an automatic method to calculate the threshold, which can be seen in Algorithm 5.2. The best value for the number of votes is estimated from its effect on the training set, but in order to alleviate computational costs, they suggest to use only 10% of the training instances.

It is desirable that the selection of votes takes into account both the training error and the percentage of features retained, which are minimized to the maximum possible extent, by minimizing the fitness criterion $e[v]$:

$$e[v] = \alpha \times error + (1 - \alpha) \times feat Percentage \qquad (5.1)$$

where α is a parameter in $[0,1]$ which measures the relative relevance of both values and v is any possible value for the threshold. Notice that the maximum number of votes is the number of weak selectors we have.

Algorithm 5.2: Pseudocode to use classification accuracy and percentage of selected features to join subsets of features

---

Data: $D_{(m \times s)}$ = training dataset with m samples and s features

       n = number of weak selectors
Result: S = final subset of features
1 **for** $i = 1$ *to* n **do**
2 | S_i = subset of features selected by the ith weak selector
3 | increment one vote for each feature not in S_i
  end
  /* Obtain threshold of votes, Th, to remove a feature   */

4 $minVote$ = minimum threshold considered (1)
5 $maxVote$ = maximum threshold considered (number of weak selectors, n)
6 $\mathbf{z}$ = submatrix of $\mathbf{D}$ with only 10% of samples
7 **for** $v = minVote$ *to* $maxVote$ **do**
8 | F_{th} = subset of selected features (number of votes $< v$)
9 | $error$ = classification error after training $\mathbf{z}$ using only features in F_{th}
10 | $featPercentage$ = percentage of features retained $\left(\frac{|F_{th}|}{|X|} \times 100 \right)$
11 | $e[v] = \alpha \times error + (1 - \alpha) \times featPercentage$
  end
12 $Th = min(e)$, Th is the value which minimizes the error e
13 S = subset of features after removing all features with a number of votes $\geq Th$

5.2.3 Using Complexity Measures

The problem with the previous approach is that using classification to combine the subsets of features implies a high computational cost, as well as being dependent on the classifier chosen. In some cases, it is even possible that the required time for this task is higher than the time necessary for the feature selection process.

Trying to overcome these issues, Morán-Fernández et al. [8] proposed to modify the function for calculating the threshold of votes presented in the previous subsection by making use of data complexity measures [7]. The reason for this decision was that they assume that good candidate features would contribute to decrease the theoretical complexity of the data and must be maintained. In order to have a methodology independent of the classifier and applicable to both binary and multiclass datasets, they chose the Fisher's multiple discriminant ratio for C classes:

$$f = \frac{\sum_{i=1, j=1, i \neq j}^{C} p_i p_j (\mu_i - \mu_j)^2}{\sum_{i=1}^{C} p_i \sigma_i^2},$$

where μ_i, σ_i, p_i are the mean, variance and proportion of the ith class, respectively. The inverse of the Fisher ratio is used, $1/f$ —from now on noted as $F1$— where a small complexity value represents an easier problem. Therefore, the new formula for calculating $e[v]$ is defined as:

$$e[v] = \alpha \times F1 + (1 - \alpha) \times featPercentage \tag{5.2}$$

The pseudocode to combine subsets of features which uses complexity measures can be seen in Algorithm 5.3.

Algorithm 5.3: Pseudocode to use complexity measures and percentage of selected features to join subsets of features

---

Data: $D_{(m \times s)}$ = training dataset with m samples and s features

       n = number of weak selectors

Result: S = final subset of features

1 **for** $i = 1$ *to* n **do**

2   S_i = subset of features selected by the ith weak selector

3   increment one vote for each feature not in S_i

   end

   /* Obtain threshold of votes, Th, to remove a feature   */

4 $minVote$ = minimum threshold considered (1)

5 $maxVote$ = maximum threshold considered (number of weak selectors, n)

6 z = submatrix of **D** with only 10% of samples

7 **for** $v = minVote$ *to* $maxVote$ **do**

8   F_{th} = subset of selected features (number of votes $< v$)

9    $complexityMeasure$ = value of F1 computed on training dataset

10    $featPercentage$ = percentage of features retained $\left(\frac{|F_{th}|}{|X|} \times 100 \right)$

11    $e[v] = \alpha \times F1 + (1 - \alpha) \times featPercentage$

   end

12 $Th = min(e)$, Th is the value which minimizes the error e

13 S = subset of features after removing all features with a number of votes $\geq Th$

5.3   Combination of Rankings of Features

In the previous section, we have seen how to combine the results obtained by the weak selectors when their output is a subset of features. But, as seen in Chap. 2, there are feature selection methods that return an ordered ranking of all the features, according to their relevance. In this case, it is necessary to find methods that can receive as an input several ranking obtained by the different weak selectors and combine them into a single final ranking, trying not to incur in an important loss of information.

Let us suppose that we have a dataset with a number of features m that will be ranked by different weak selectors. Depending on the ensemble strategy, it is possible that all the weak selectors have rank all the features, or only a subset of them. For this example, let us assume the worst case scenario, which is when weak selectors work with subsets of features, so the rankings they produce are called partial or incomplete rankings.

To illustrate this problem, let n be the number of weak or base selectors conforming the ensemble, while the relevance of each feature is randomly generated as a number between 0 and 1. Once the partial rankings (obtained from each weak selector) are combined somehow, we would need a measure to determine to what extent the *combined* ranking is close to the *ideal* ranking (obtained when working with all data). For this task, we can use the Normalized Discounted Cumulative Gain (NDCG) [9], which is often used to measure effectiveness of web search engine algorithms or related applications. This method returns a value between 0 and 1, where 1 means that the rankings are identical. The pseudocode for this toy example can be found in Algorithm 5.4.

Algorithm 5.4: Pseudo-code for generating the toy example

Data: $D_{(m \times s)}$ = training dataset with m samples and s input features
  n = number of weak selectors
  $\mathbf{X}$ = set of features, $\mathbf{X} = \{X_1, \ldots, X_s\}$
  s_n = number of features to go to each weak selector

Result: $NDCG$ = similarity between the true ranking and the combined ranking

1 Generate a random value between 0 and 1, $Score(X_i)$, for each feature $X_i \in \mathbf{X}$, obtaining a true ranking $Rank_t$
2 **for** $i = 1$ *to* n **do**
3 $\quad$ $D_{(m \times s_i)}$ = subset of data with s_n random features
4 $\quad$ Rank the features according to their $Score$, obtaining a partial ranking $Rank_p(i)$
  end
5 **for** *each feature s in* $\mathbf{X}$ **do**
6 $\quad$ $Avg(s)$ = calculate the average of its position in all the partial rankings $Rank_p(i), \forall i \in n$
  end
7 Obtain a combined ranking $Rank_c$ by ordering Avg
8 $NDCG$ = compare $(Rank_t, Rank_c)$

Figure 5.3 shows an example in which the number of features to rank is $s = 100$ and the maximum number of weak selectors available is $n = 100$. The NDCG value is represented by the color, as the colorbar in the right side of the figure depicts. As expected, when all the features are ranked by each weak selector, the NDCG value is 1 since the rankings are identical. Nevertheless, even when we have 10 weak selectors and 80 features ranked by each weak selector, the rankings are not exactly the same, which gives us an idea about the complexity of the ranking combination task. From the figure, we can see that, for ensuring good results to be obtained, it is necessary to work with complete rankings, but this is not always possible. And notice that this example does not even reflect what happens in a real situation, in which a

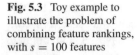

Fig. 5.3 Toy example to illustrate the problem of combining feature rankings, with $s = 100$ features

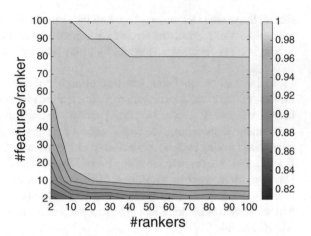

weak selector has only partial information to rank the features, so the importance given to each feature cannot be computed in an accurate way. On the contrary, in this example, for each weak selector we are using the *true* importance of each feature.

As can be seen through this simple example, the problem of combining rankings is not trivial, even when we can use the true importance of the features, obtained from the whole set of examples. Arrow's impossibility theorem [10] states that, when having *at least* two rankers and *at least* three options to rank (in this case features), it is impossible to design an aggregation function that satisfies in a strong way a set of desirable conditions at once, such that:

- If every weak selector ranks feature X over feature Y, then the final ranking has X over Y.
- If every weak selector's preference between X and Y remains unchanged, then the final ranking's preference between X and Y will also remain unchanged (even if weak selectors' preferences between other pairs like X and Z, Y and Z, or Z and W change).
- There is no 'dictator': no single weak selector possesses the power to always determine the final ranking's preference.

So, this theorem also acknowledges how challenging it is to combine partial rankings. However, there are cases in which the loss of information suffered from combining partial rankings does not reflect on important loss of subsequent classification accuracy [11] and so researchers are still using ensemble approaches or other methods that require to combine partial rankings. The remainder of this section describes the most popular aggregation methods to combine rankings. In Chap. 6, Sect. 6.2, we can see an example of the different behaviors shown by these combination methods within the implementation of an ensemble for feature selection.

5.3.1 Simple Operations Between Ranks

The easiest way to combine rankings of features is to apply simple operations through them, such as the median or the mean. Some popular methods can be defined as follows:

- **min**: assigning to each element to be ranked the minimum (best) position that it has achieved among all rankings.
- **median**: assigning to each element to be ranked the median of all the positions that it has achieved among all rankings.
- **arith.mean**: assigning to each element to be ranked the mean of all the positions that it has achieved among all rankings.
- **geom.mean**: assigning to each element to be ranked the geometric mean of all the positions that it has achieved among all rankings.

To illustrate the behavior of these methods, we will use a simple example. Suppose that the have five features to be ranked $\{f1, f2, f3, f4, f5\}$, and 5 different rankings of them $R_1, R_2, \ldots, R_5$, as depicted in Table 5.1. The last rows of the table show the calculations that each method has to do. Notice that the method 'min' computes the best value achieved by each feature along the different rankings ('best' meaning the highest position). In this case, there was a tie, and according to the implementation provided by [12] the method returns the elements which are tied in their original position. Thus, for this example, all the methods return the final ranking $\{f1, f2, f3, f4, f5\}$.

Suppose now that we are working with an ensemble which distributes the data, such that it is possible that each weak feature selector does not have access to all the features. As mentioned before, the rankings obtained by each weak selector are called partial or incomplete rankings. Imagine that we have six features to be ranked $\{f1, f2, f3, f4, f5, f6\}$ and 3 weak selectors, provided that three features go to each weak selector so as to guarantee some overlap between the different partial rankings. Then, we may have, for instance, features $\{f1, f2, f3\}$ for the first weak selector, features $\{f3, f4, f6\}$ for the second one, and features $\{f1, f5, f6\}$ for the third one. The three partial rankings can be seen in Table 5.2, in which features that are not present for a given weak selector are being assigned the last position in the

Table 5.1 Example of aggregation methods which use simple operations

Feature	R_1	R_2	R_3	R_4	R_5	min	median	arith mean	geom mean
$f1$	1	2	3	1	1	1	1	1.6	1.4
$f2$	2	1	1	2	3	1	2	1.8	1.6
$f3$	3	3	2	5	2	2	3	3.0	2.8
$f4$	4	4	5	3	4	3	4	4.0	3.9
$f5$	5	5	4	4	5	4	5	4.6	4.6

Table 5.2 Example of aggregation methods which use simple operations working with partial rankings

Element	R_1	R_2	R_3	min	median	arith mean	geom mean
$f1$	1	6	1	1	1	2.7	1.8
$f2$	3	6	6	3	6	5.0	4.8
$f3$	2	3	6	2	3	3.7	3.3
$f4$	6	1	6	1	6	4.3	3.3
$f5$	6	6	3	3	6	5.0	4.8
$f6$	6	2	2	2	2	3.3	2.9

ranking, according to the implementation provided by [12] (6, in this example). In this case, the *min* method will return $\{a, d, c, f, b, e\}$, whereas the remaining methods will return $\{f1, f6, f3, f4, f2, f5\}$. Notice that, for the sake of this example, we are choosing the alphabetical order in case of ties.

Some of these methods are more likely to have to deal with ties than others (e.g. *min* and *median* methods are prone to have ties, since the set of possible values obtained by them are much more reduced than the possible values received by *arith mean* or *geom mean*). However, in a previous work [11], we have demonstrated the superiority of the *min* method on a set of microarray datasets, since it was the only method that was able to overlook the presence of many 'last' positions for each feature, which greatly affected the performance of the other methods.

5.3.2 Stuart Aggregation Method

Stuart et al. [13] introduced the first attempt to use order statistics for combination of rankings, although the computational scheme for their method was further optimized by Aerts et al. [14]. This method compares the actual rankings with the expected behavior of uncorrelated rankings, re-ranks the features and assigns significance scores. Despite being robust to noise, this method requires simulations to define significance thresholds and does not support partial rankings.

5.3.3 Robust Rank Aggregation

This method was proposed by Kolde et al. [12] to improve the limitations of Stuart and classical methods. Their authors state that this algorithm is both computationally efficient and statistically stable. With this method, the combination is based on the comparison of the actual data with a null model that assumes random order of the weak rankings. A P-value assigned to each feature in the aggregated ranking described how much better it was ranked than expected. This provides basis for reordering and identifies significant features. As the P-value calculation procedure takes into account only the best ranks for each feature, the method is said to be very robust.

5.3.4 SVM-Rank

SVM-Rank [15] is an *SVM*-based method that can be trained to learn ranking functions. The *SVM-Rank* algorithm considers a training set S of size n containing feature selection methods q with their rankings r according to (5.3):

$$(q_1, r_1), (q_2, r_2), \ldots, (q_n, r_n) \tag{5.3}$$

The algorithm selects a ranking function f that maximizes Eq. (5.4):

$$\tau_S(f) = \frac{1}{n} \sum_{i=1}^{n} \tau(r_{f(q_i)}, r_i) \tag{5.4}$$

The function f must maximize Eq. (5.4) and must generalize beyond the training data. Consider the class of linear ranking functions (5.5) defined as:

$$(c_i, c_j) \in f_{\vec{w}}(q) \Leftrightarrow \vec{w} \, \Phi(q, c_i) > \vec{w} \, \Phi(q, c_j), \tag{5.5}$$

where $\vec{w}$ is a weight vector that is adjusted by learning. $\Phi(q, c)$ is a mapping between method q and feature c. For any weight vector $\vec{w}$, the points are ordered by their projection onto $\vec{w}$. Maximizing (5.4) is equivalent to finding the weight vector so that the maximum number of the following inequalities is satisfied (5.6):

$$\forall(c_i, c_j) \in r_k : \vec{w} \, \Phi(q_k, c_i) > \vec{w} \, \Phi(q_k, c_j) \mid k = 1, \ldots, n \tag{5.6}$$

The solution to this problem is approximated, analogously to *SVM* classification, by introducing slack variables $\xi_{i,j,k}$ and minimizing the upper bound $\sum_{i,j,k}$. This renders the problem equivalent to an *SVM* classification on pairwise difference vectors $\Phi(q_k, c_i) - \Phi(q_k, c_j)$.

5.4 Summary

In this chapter we have described the different methods available for combining the weak outputs obtained from the ensemble. We started with the case in which we combine the outputs of the classifiers, and then we moved to techniques to combine subsets of features and rankings of features. The combination of partial outputs is a crucial point in the design of an ensemble and must be taken into account with care, since the final output will depend on the election made.

References

1. Kuncheva, L.I.: Combining Pattern Classifiers: Methods and Algorithms. Wiley, New Jersey (2004)
2. Peteiro-Barral, D., Guijarro-Berdiñas, B.: A survey of methods for distributed machine learning. Prog. Artif. Intell. **2**(1), 1–11 (2013)
3. Álvarez-Estévez, D., Sánchez-Maroño, N., Alonso-Betanzos, A., Moret-Bonillo, V.: A survey of methods for distributed machine learning. Expert Syst. Appl. **38**(6), 7746–7754 (2011)
4. Aguilar-Ruiz, J.S., Azuaje, F., Riquelme, J.C.: Data mining approaches to diffuse large B-Cell lymphoma gene expression data interpretation. Lecture Notes in Computer Science, pp. 279–288. Springer, Berlin (2004)
5. Bolón-Canedo, V., Sánchez-Maroño, N., Alonso-Betanzos, A.: Distributed feature selection: an application to microarray data classification. Appl. Soft Comput. **30**, 136–150 (2015)
6. Bolón-Canedo, V., Sánchez-Maroño, N., Cerviño-Rabuñal, J.: Toward parallel feature selection from vertically partitioned data. In: Proceedings of European Symposium on Artificial Neural Networks, ESANN, pp. 395–400 (2014)
7. Basu, M., Ho, T.K.: Data Complexity in Pattern Recognition. Springer, Berlin (2006)
8. Morán-Fernández, L., Bolón-Canedo, V., Alonso-Betanzos, A.: Centralized versus distributed feature selection methods based on data complexity measures. Knowl. Based Syst. **117**, 27–45 (2017)
9. Järvelin, K., Kekäläinen, J.: Cumulated gain-based evaluation of IR techniques. ACM Trans. Inf. Syst. **20**(4), 422–446 (2002)
10. Arrow, K.J.: Social Choice and Individual Values. Wiley, New Jersey (1951)
11. Bolón-Canedo, V., Sechidis, K., Sánchez-Maroño, N., Alonso-Betanzos, A., Brown, G.: Exploring the consequences of distributed feature selection in DNA microarray data. In: Proceedings of international joint conference on neural networks, IJCNN, pp. 1665–1672 (2017)
12. Kolde, R., Laur, S., Adler, P., Vilo, J.: Robust rank aggregation for gene list integration and meta-analysis. Bioinformatics **28**(4), 573–580 (2012)
13. Stuart, J., Segal, E., Koller, D., Kim, S.K.: A gene-coexpression network for global discovery of conserved genetic modules. Science **302**(5643), 249–255 (2003)
14. Aerts, S., Lambrechts, D., Maity, S., Van Loo, P., Coessens, B., Tranchevent, L.C., De Moor, B., Marynen, P., Hassan, B., Carmeliet, P., et al.: Gene prioritization through genomic data fusion. Nat. Biotechnol. **24**(5), 537–544 (2006)
15. Joachims, T.: Optimizing search engines using clickthrough data. In: Proceedings of the eighth ACM SIGKDD international conference on knowledge discovery and data mining, pp. 133–142 (2002)

Chapter 6
Evaluation of Ensembles for Feature Selection

Abstract This chapter describes the different approaches that can be used to evaluate the behavior of the ensembles for feature selection. Beside the well-known, almost universal measures of accuracy, there are two other measures that should be taken into account to quantify the success of an ensemble approach: diversity and stability. In both cases, the relation between the three measures has been studied relatively well in the field of classification ensembles. However, the situation is quite different in the case of ensembles for feature selection, in which measures for diversity and stability have not been devised specifically and more research and proposals are needed. Section 6.1 states the basic ideas on the evaluation of ensembles. Then, Sect. 6.2 defines the concept of *diversity* and describes some recent attempts in evaluating diversity and in using it as a measure to be balanced with accuracy in order to devise more powerful ensembles. Section 6.3 comments on the stability of feature selection ensembles and Sect. 6.4 defines performance evaluation measures for both subsets of features and rankings of features. Finally, Sect. 6.5 summarizes and discusses the contents of this chapter.

Evaluation of feature selection ensembles is still a scarce reference in scientific literature. While performance is of course a universal measure, diversity and stability are also factors that have relevance in the process, as on the one hand we need to ensemble single methods that produce diverse results but also on the other hand we need robust ensembles. So far, and although measures for diversity and stability in classifier ensembles have been devised, the subjects are still rare for the case of feature selection ensembles. In this chapter, we will discuss some basic ideas and experimental results.

6.1 Introduction

Boost in accuracy is the crucial reason for using ensembles in machine learning. But in the evaluation of ensembles, there are other two important parameters involved, diversity and robustness. Diversity is at the core of the basic idea of devising ensemble

© Springer International Publishing AG, part of Springer Nature 2018
V. Bolón-Canedo and A. Alonso-Betanzos, *Recent Advances in Ensembles
for Feature Selection*, Intelligent Systems Reference Library 147,
https://doi.org/10.1007/978-3-319-90080-3_6

methods, as in the case of classification the examples which are misclassified by some members of the ensemble are correctly classified by others, in such a way that the final accuracy is greater than it is with any of the single classifiers. Thus, diversity–or disagreement within the ensemble– among the members of the ensemble is a key issue in the combination of single classifiers [1, 2]. In several studies it is shown that the use of diversity positively affects the quality of classification [3, 4]. Considering diversity and accuracy simultaneously has been also an strategy for ensemble pruning, aiming at obtaining better generalization capabilities in classification, for example [5].

There are several statistics that can be used as a measure of diversity. In the article by Kuncheva and Whitaker [6] the pair-wise Q statistics [7] is recommended, as it is simple to understand and to implement. Although there are several works regarding diversity in ensembles for classification [3, 6, 8, 9], there is a necessity for the establishment of novel diversity measures for ensembles for other machine learning algorithms, as feature selection or one-class classification [10, 11]. Not only diversity is important, but also the function that combines the results of the different components of the ensemble (see Chap. 5 for the description of several methods). As early as in the work described in [12], it was shown that diversity in the feature subset created alone is not enough for increasing the accuracy of the machine learning process, as the combination method should also make proper use of the diversity obtained in order to maintain the benefit.

Another important factor for the evaluation of ensembles for feature selection is stability or robustness, that is, the capacity that ensembles have for returning an stable subset or rank of features, as in individual feature selection methods the process might be unstable, and depend on the portion of the training set used. There are several measures that have been used to quantify stability in FS processes [13], depending mainly on the type of output of the methods employed (ranking, weighting or feature subset). In the mentioned study by Brown and Nogueira, they enumerate the properties that the measure should have, and suggest as conclusion that the well-known Pearson correlation is the most adequate similarity measure for robustness. But not only measuring robustness is important, but also how we can make feature selection procedures more stable. At this respect, there are several experimental studies [14–18] that have shown that ensemble approaches overcome standard selection algorithms in terms of stability, especially in the context of high-dimensional/small sample size domains (such as microarrays, genomics, etc.). In [14], the authors have analyzed both accuracy and stability of different implementations, concluding that the benefit achieved by the ensemble approach increases when the strength of the individual method decreases, that is, the effect of the ensemble strategy is to approach the results of the weakest and strongest methods, leading to more stable and accurate behaviors.

6.2 Diversity

Ensemble feature selection is one of the strategies that, by incorporating diversity, aims to obtain an optimal feature subset or feature ranking. Although, as detailed above, there have been several measures that can quantify diversity in ensembles of classifiers or regressors [6, 8], very little effort has been done in studying their adequacy to ensemble feature selection. In [10], several experiments have been conducted in order to test the use of such diversity measures over a suite of 21 datasets. Their conclusions are that the performance of the ensemble might be influenced by the diversity measure chosen, and of course, also on the type of dataset being processed. Their final recommendation is that in most cases, the plain disagreement measure is the one that behaves best.

In [19] simple random selection of feature subsets (named Random Subspacing–RS–) is used for introducing diversity. The idea is to randomly select a number F^* of features from the F-dimensional training set, repeating the process S times to build S feature subsets employed to construct S base classifiers. In [10] the authors employed probabilistic feature selection instead of choosing a fixed number of F features. This implementation showed ensembles that obtained higher diversity and accuracy. The research described in this article is one of the very few studies in diversity for feature selection ensembles. In the article the authors compare five measures of diversity using a wrapper approach for the feature selection ensembles in which several search strategies are employed. In [15], only feature selection rankers are used, three filter (Information Gain, Minimum Redundancy Maximum Relevance-mRMR- and ReliefF) and two embedded methods (Recursive Feature Elimination for Support Vector Machines-SVM-RFE- and Feature Selection Perceptron–FS-P). In order to ensure the diversity of results from these widely used methods, the Spearman rank coefficient and the Kendall rank correlation coefficient were used. The results of those tests over two datasets are shown in Table 6.1 for the Spearman coefficient [20]. The ρ value in the range $[-1, 1]$ reflects the relationship between rankings, with 1 indicating that the compared rankings were equal.

Similar results were obtained for the Kendall coefficient [21], as shown in Table 6.2, where it can be seen that most of the ρ values are far from 1, indicating great differences between the paired rankings (obviously, when the same ranker method rankings were compared, the ρ value was 1, as can be seen in the table diagonals). This small experiment, using only two of the datasets employed in their experimental study, (*Spambase* and *Isolet*), aimed at demonstrating that the set of feature selection rankers chosen for this study ensured enough diversity in their behaviors.

As it was mentioned in Sect. 6.1 of this chapter, and taking into consideration the results of the work described in [12], diversity in the feature subsets is not enough as the combination method should make proper use of the diversity created in order to maintain the benefit. In the work described in [15] the authors employed several combination methods (also called aggregators), to test whether their selection influences on the final results obtained after classification. The aggregators tested were the ones

Table 6.1 ρ value of Spearman's rank correlation coefficient

Dataset	Ranker	InfoGain	mRMR	ReliefF	SVM-RFE	FS-P
Spambase	InfoGain	1.0000	0.2011	0.0714	−0.2040	−0.1736
	mRMR	0.2011	1.0000	−0.0811	0.1313	0.0838
	ReliefF	0.0714	−0.0811	1.0000	−0.0672	0.0380
	SVM-RFE	−0.2040	0.1313	−0.0672	1.0000	0.0565
	FS-P	−0.1736	0.0838	0.0380	0.0565	1.0000
Isolet	InfoGain	1.0000	0.0971	−0.0677	−0.0320	−0.0521
	mRMR	0.0971	1.0000	0.0295	0.0534	0.0062
	ReliefF	−0.0677	0.0295	1.0000	0.0115	−0.0291
	SVM-RFE	−0.0320	0.0534	0.0115	1.0000	0.0331
	FS-P	−0.0521	0.0062	−0.0291	0.0331	1.0000

Table 6.2 ρ value of Kendall's rank correlation coefficient

Dataset	Ranker	InfoGain	mRMR	ReliefF	SVM-RFE	FS-P
Spambase	InfoGain	1.0000	0.1278	0.0476	−0.1466	−0.1266
	mRMR	0.1278	1.0000	−0.0602	0.0940	0.0464
	ReliefF	0.0476	−0.0602	1.0000	−0.0489	0.0288
	SVM-RFE	−0.1466	0.0940	−0.0489	1.0000	0.0351
	FS-P	−0.1266	0.0464	0.0288	0.0351	1.0000
Isolet	InfoGain	1.0000	0.0652	−0.0449	−0.0212	−0.0337
	mRMR	0.0652	1.0000	0.0216	0.0373	0.0053
	ReliefF	−0.0449	0.0216	1.0000	0.0084	−0.0168
	SVM-RFE	−0.0212	0.0373	0.0084	1.0000	0.0213
	FS-P	−0.0337	0.0053	−0.0168	0.0213	1.0000

listed in Table 6.3, that are included in the *RobustRankAggreg* package implemented in the R and Matlab languages [22]. These combination methods consider a set of n feature selection ranker methods, where $\mathcal{Q} = \{q_i, i = 1, \ldots, n\}$, and where each q_i is associated with a list of m objects that represent the relevance of features in the range [0, 1]. Once the relevance of each feature in the individual ranking method is obtained, one of the reduction functions shown in Table 6.3 is applied (see also Chap. 5). The result is a reduced final ranking that is ordered according to the calculated relevance factor. The final relevance values of the features will be in the range [0, 1], where higher and lower values reflect more and less important features, respectively, in the dataset.

In the work described in [15], SVM classifier has been used to assess the final performance of the ensemble, and thus a specific aggregator was also tested, named SVM-Rank [24]. For more details on the combination methods, please check Chap. 5.

Using the same datasets referenced in Chap. 4, Table 4.1, the results obtained in terms of average percentage errors are shown in Fig. 6.1.

Table 6.3 Reduction functions for feature rankings

Function	Formula	Description
min	$\min\{q_1(d_j), q_2(d_j) \ldots q_n(d_j)\}$	Reduction function based on simple arithmetic operations. It selects the minimum of the relevance values yielded by the rankings [23]
median	$\mathrm{median}\{q_1(d_j), q_2(d_j) \ldots q_n(d_j)\}$	Reduction function based on simple arithmetic operations. It selects the median of the relevance values yielded by the rankings [23]
mean	$\frac{1}{n}\sum_{i=1}^{n} q_i(d_j)$	Reduction function based on simple arithmetic operations. It selects the average of the relevance values yielded by the rankings [23]
geomMean	$\left(\prod_{i=1}^{n} q_i(d_j)\right)^{1/n}$	Reduction function based on simple arithmetic operations. It selects the geometric average of the relevance values yielded by the rankings [23]
Stuart	$Pq[X \leq \rho] = 1 - Pq[\hat{q}_1 \leq 1 - \mathscr{B}_{n,n}^{-1}(\rho), \ldots, \hat{q}_n \leq 1 - \mathscr{B}_{n,1}^{-1}(\rho)]$	Reduction function based on statistical sorting distributions. It uses the *Beta* distribution to obtain the ρ value [21]
RRA	$\min_{i=1,\ldots,n} \mathscr{B}_{k,n}(r), \quad \mathscr{B}_{k,n}(q) = Pr[\hat{q}_k \leq q_k]$	Reduction function based on statistical sorting distributions. Based on the *Stuart* function, it improves the efficiency-accuracy connection through the use of *Bonferroni* correction when calculating the ρ value [22]

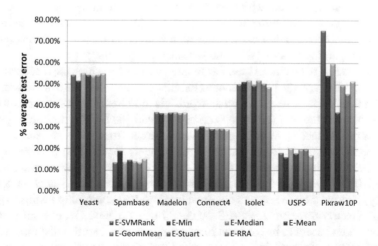

Fig. 6.1 Comparison of average estimated percentage test errors for the different combination methods

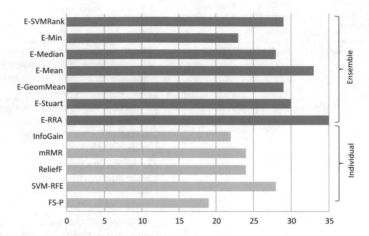

Fig. 6.2 Number of cases when the results obtained by the individual and the heterogeneous centralized ensemble approaches were comparable with the best result

As can be seen, the different combination methods obtained similar results except the ensemble that used the *Min* reduction function (second bar in each set of bars), that produced irregular results (the best average test error for the *Yeast*, *Madelon* and *USPS* datasets, but the worst average test error for the *Spambase* and *Connect4* datasets). The case of the dataset Pixraw10P is different from the other, as it is a microarray dataset, in which the number of features is much higher than the number of samples. In this case, accuracy varied greatly depending on the combination method, with *E-SVMRank* achieving the worst result, and *E-Mean* obtaining the best result. The choice of the combination method appears thus to have a great influence in the final results for microarray datasets—a conclusion which is consistent with that reported elsewhere regarding an extensive study of microarray datasets [25].

Finally, in Fig. 6.2 it can be seen the number of cases for which the results obtained by the individual and the heterogeneous ensemble approaches were not significantly different than the best result (in other words, the number of times that results were comparable with the best result). As can be observed, the *E-RRA* ensemble approach obtained results that were not significantly different from the best result in all 35 experiments, compared to 28 out of 35 experiments for the best performing individual feature selection method (*SVM-RFE*). Five of the remaining six ensemble methods (*E-SVMRank*, *E-Median*, *E-Mean*, *E-GeomMean* and *E-Stuart*) matched or (mostly) improved on the results obtained by *SVM-RFE*, obtaining results that were not significantly different in 28–33 of the 35 experiments. Overall, an ensemble approach would seem to be the most reliable approach to feature selection, although in some specific cases, an individual method (not always the same one) might well perform better than the ensemble. Besides, the results show than the choice of combination method influences the performance results, and that the RRA aggregator is, on average, the one that obtains the best results.

6.3 Stability

As mentioned in the previous section, it is desirable that the methods chosen for conforming the ensemble are *diverse*, i.e. that they provide different enough outputs on the same sample of data. However, when we are changing the sample of data, it is desirable that such methods return similar outputs, which is known as *stability*. Thus, the stability of a feature selection method can be seen as its sensitivity to small changes in the input dataset.

As pointed out by Nogueira and Brown [26], in ensemble-based feature selection, the goal must be to use diverse feature selection methods within the ensemble (corresponding to low stability), as well as obtaining robustness of the final feature selection made by the ensemble (corresponding to high stability). Therefore, the stability of ensembles for feature selection has been gaining attention in recent years [27–30].

There are plenty of measures in the literature to compute stability, and in the following we will comment on the most popular ones, according to if they are designed for subsets of features or for rankings of features.

6.3.1 Stability of Subsets of Features

This is the most common approach in the literature, since even in the case of rankers, it is possible to establish a threshold and compare the top k selected features. Two of the first methods to measure similarity are Jaccard index [31] (also referred as Tanimoto distance) or the relative Hamming distance [32]:

$$Jac(A, B) = \frac{|A \cap B|}{|A \cup B|} = \frac{|A \cap B|}{|A| + |B| + |A \cap B|}. \tag{6.1}$$

$$Ham(A, B) = 1 - \frac{|A \setminus B| + |B \setminus A|}{n}. \tag{6.2}$$

However, both these measures are subset-size-biased [33], which means that they provide different results depending on the number of features selected so they cannot be considered consistent. Suppose that a feature selection procedure selects two identical feature sets of eight features out of a total of 10 features, and another procedure selects also two identical feature sets of eight features but, in this case, out of a total of 100 features. Intuitively, we can see that the second procedure is more stable, but the two measures defined above would give us the same result. For this reason, Kuncheva [33] identified the correction for chance as one of the three desirable properties for a stability measure. Let A and B be subsets of features, of the same cardinality k. Let $r = |A \cap B|$ be the cardinality of the intersection of the two subsets. Then, the properties are:

- *Monotonicity.* For a fixed cardinality of the subsets k, and for a number of features n, the larger the intersection between subsets, the higher the value of the stability measure.
- *Limits.* A stability measure should be bound by constants not dependent on k or n. The maximum stability should be reached when the two subsets are identical (i.e. when $r = k$).
- *Correction for chance.* A stability measure should have a constant value for independently drawn subsets of features of the same cardinality.

Kuncheva [33] proposed a consistency index to measure stability that satisfies the three properties described above:

$$Kun(A, B) = \frac{f - \frac{k^2}{n}}{k - \frac{k^2}{n}} = \frac{rn - k^2}{k(n - k)}, \tag{6.3}$$

such that $|A| = |B| = k$ and where $0 < k < |X| = n$.

However, a problem with this stability measure is that it requires that subset sizes are the same, which in practice does not always happen. Therefore, Nogueira and Brown [26] added to this one another two desirable properties for a stability measure to have:

- *Unconstrained on cardinality.* A stability measure should be able to deal with feature sets of different cardinalities.
- *Symmetry.* A stability measure should be symmetrical, so that its value does not depend on the order on which the feature sets are taken.
- *Redundancy awareness.* Since feature can be redundant among each other, a stability measure should take this fact into account.

Nogueira and Brown [26] provided a summary of the most commonly used stability measures pointing out if they have these six desirable properties, as depicted in Table 6.4. Notice that the measures proposed by Lustgarten [34], Wald [35] and Zhang [36] are all variants of Kuncheva's similarity measure for feature sets of varying cardinalities.

6.3.2 Stability of Rankings of Features

As we have seen in Chap. 2, there are feature selection methods that return an ordered ranking of all the features, according to their relevance. In this case, we cannot use the stability measures mentioned in the previous subsection, unless we decide to consider only the top k features, but even in this case it would be not absolutely correct because we would be losing the order/relevance of the features.

Among the most popular measures to compute the similarity between rankings we can find the Kendall Tau [40], the Canberra Distance [41] and the Spearman's

Table 6.4 Properties of popular stability measures

	Monotonicity	Limits	Correction	Cardinality	Symmetry	Redundancy
Jaccard [32]	✓	✓		✓	✓	
Hamming [31]	✓	✓		✓	✓	
Yu [37]	✓	✓		✓	✓	✓
Kuncheva [33]	✓	✓	✓		✓	
Lustgarten [34]		✓	✓	✓	✓	
Wald [35]		✓	✓	✓	✓	
nPOG [36]		✓	✓	✓		
nPOGR [36]		✓	✓	✓		✓
CW_{rel} [38]	✓	✓		✓	✓	
Krízek [39]	✓	✓			✓	

ρ [31]. Let R_1 and R_2 be two rankings and f the number of features in the dataset, these measures can be defined as follows:

$$Spear(R_1, R_2) = 1 - \frac{6\sum d^2}{f(f^2 - 1)},$$ (6.4)

where d is the distance between the same feature in both rankings.

$$Cam(R_1, R_2) = \sum_{i=1}^{f} \frac{|R_{1_i} - R_{2_i}|}{|R_{1_i}| + |R_{2_i}|}$$ (6.5)

$$Kend(R_1, R_2) = \sum_{\{i,j\}\in P} \bar{K}_{i,j}(R_1, R_2)$$ (6.6)

where

P is the set of unordered pairs of distinct elements in R_1 and R_2
$\bar{K}_{i,j}(R_1, R_2) = 0$ if i and j are in the same order in R_1 and R_2
$\bar{K}_{i,j}(R_1, R_2) = 1$ if i and j are in the opposite order in R_1 and R_2.

As mentioned before, it is desirable that the feature selection methods chosen for conforming the ensemble are stable to changes in the training set. Moreover, it is a common belief that by combining a set of unstable individual feature selectors and aggregating them together in an ensemble would increase stability. Nogueira et al. [42] focused on the case of combining the individual output rankings using mean rank aggregation, and demonstrated that the error of the aggregated rankings is guaranteed to be lower than the one of an individual ranking on average. Moreover, they gave a theoretical argument showing why the stability of the aggregated rank improves as the number of ensemble member increases.

6.4   Performance of Ensembles

When evaluating the quality of an ensemble of feature selection methods, it is common—and desirable—to employ the measures described above, i.e. the diversity between the weak selectors and the stability and robustness of them. But, eventually, an ensemble for feature selection has be to evaluated by its performance, which can be its ability to select the relevant features (only possible when we know what the relevant features are), or the classification accuracy obtained with the selected features.

6.4.1   Are the Selected Features the Relevant Ones?

In an ideal situation, it would be perfect to be able to evaluate a feature selection system based only on the quality of the features selected, without involving any classifier. But, in practice, the set of relevant features are not known a priori unless we are using artificial data. In fact, several authors choose to use artificial data stating that although the final goal of a feature selection method is to test its effectiveness over a real dataset, the first step should be on synthetic data. The reason for this is two-fold [43]:

1. Controlled experiments can be developed by systematically varying chosen experimental conditions, like adding more irrelevant features or noise in the input. This fact facilitates to draw more useful conclusions and to test the strengths and weaknesses of the existing algorithms.
2. The main advantage of artificial scenarios is the knowledge of the set of optimal features that must be selected, thus the degree of closeness to any of these solutions can be assessed in a confident way.

If we use artificial data and then we know the relevant features, there are several measures we can use to evaluate the performance of the ensemble, depending on if the ensemble returns a subset of features or a ranking of features.

6.4.1.1   Subsets of Features

Of course, the perfect behavior for a ensemble which returns a subset of features is to select only the relevant features and none of the irrelevant or redundant ones. However, this situation does not always happen, so there is a need to design measures that attempt to reward the selection of relevant features at the same time that penalize the inclusion of irrelevant ones, considering two undesirable situations:

- The solution is *incomplete*: there are relevant features lacking.
- The solution is *incorrect*: there are some irrelevant features.

It is desirable that the measures to evaluate the correct selection of the features might take into account that choosing an irrelevant feature is better than missing a relevant one (i.e. we prefer an incorrect solution rather than an incomplete one).

In the following, we describe some popular measures to evaluate the quality of subsets of selected features, provided that we know a priori the relevant ones [44]. For the description of the methods, note that *feat_sel* stands for the subset of selected features, *feats* is the total set of features, *feat_rel* is the subset of relevant features, and *feat_irr* represents the subset of irrelevant features (the last two known a priori).

- The *Hamming_loss* (H) measure evaluates how many times a feature is misclassified (selected when is irrelevant or not selected when is relevant)

$$H = \frac{\#(feat\_sel \cap feat\_irr) + \#(feat\_not\_sel \cap feat\_rel)}{\#(feat\_rel \cup feat\_irr)}$$

- The *F1-score* is defined as the harmonic mean between precision and recall. *Precision* is computed as the number of relevant features selected divided by the number of features selected; and *recall* is the number of relevant features selected divided by the total number of relevant features. Therefore, the F1-score can be interpreted as a weighted average of the precision and recall. Considered $1 - F1$-score, it reaches its best value at 0 and worst score at 1.

$$F1 = 2 \times \frac{\text{precision} \times \text{recall}}{\text{precision} + \text{recall}}.$$

6.4.1.2 Rankings of Features

In the case of ensembles that return a ranking of all the features, the measures described above are not useful because all the features are present in the ranking. A possible solution is to establish a threshold and transform the ranking in a subset of features. But there are also methods specifically defined to evaluate rankings, which in essence check if the relevant features are ranked above the irrelevant ones. Below we describe some popular ones [44]:

- The *ranking_loss* (R) evaluates the number of irrelevant features that are better ranked than the relevant ones. The fewer irrelevant features are on the top of the ranking, the best classified are the relevant ones. Notice that *pos* stands for the position of the last relevant feature in the ranking.

$$R = \frac{pos - \#feat\_rel}{\#feats - \#feat\_rel}$$

- The *average_error* (*E*) evaluates the mean of E_i, in which $i \in feats\_sel$ and E_i is the average fraction of relevant features ranked above a particular feature i.

$$E_i = \frac{\sum_j feat\_sel(j) \in feat\_rel \cap j < i - \frac{\#feat\_rel \times (\#feat\_rel - 1)}{2}}{\#feat\_irr \times \#feat\_rel}.$$

6.4.2 The Ultimate Evaluation: Classification Performance

As mentioned before, the final goal of a feature selection method is usually to test its effectiveness over a real dataset, and since in real datasets we do not know which are the relevant features, it is necessary to use a classification algorithm[1] to evaluate the performance of the feature selection process, focusing on the classification accuracy. Unfortunately, the class prediction depends also on the classification algorithm used, so when testing a feature selection result, a common practice is to use several classifiers to obtain results as classifier-independent as possible. In the following we describe some of the most common classification algorithms. Notice that some of them only can work with categorical features, whereas others require numerical attributes. In the first case, the problem is often solved by discretizing the numerical features. In the second case, it is common to use a conversion method which assigns numerical values to the categorical features.

6.4.2.1 Support Vector Machine, SVM

A Support Vector Machine [45] is a learning algorithm typically used for classification problems (text categorization, handwritten character recognition, image classification, etc.). More formally, a support vector machine constructs a hyperplane or set of hyperplanes in a high- or infinite-dimensional space, which can be used for classification, regression, or other tasks. Intuitively, a good separation is achieved by the hyperplane that has the largest distance to the nearest training data point of any class (so-called functional margin), since in general the larger the margin the lower the generalization error of the classifier. In its basic implementation, it can only work with numerical data and binary classes.

---

[1]In fact, it is possible to use any learning algorithm, such as regression, clustering, etc., depending on the task we are dealing with. However, in this book we are focusing by default on classification, since it is the most popular learning algorithm used after feature selection.

6.4.2.2   Proximal Support Vector Machine, PSVM

This method classifies points assigning them to the closest of two parallel planes (in input or feature space) that are pushed as far apart as possible [46]. The difference with a Support Vector Machine (SVM) is that PSVM classifies points by assigning them to one of two disjoint half-spaces. The PSVM leads to an extremely fast and simple algorithm by generating a linear or nonlinear classifier that merely requires the solution of a single system of linear equations.

6.4.2.3   C4.5

C4.5 is a classifier developed by [47], as an extension of the ID3 algorithm (Iterative Dichotomiser 3). Both algorithms are based in decision trees. A decision tree classifies a pattern doing a descending filtering of it until finding a leaf, that points to the corresponding classification. One of the improvements of C4.5 with respect to ID3 is that C4.5 can deal with both numerical and symbolic data. In order to handle continuous attributes, C4.5 creates a threshold and depending on the value that takes the attribute, the set of instances is divided.

6.4.2.4   Naive Bayes, NB

A naive Bayes classifier [48] is a simple probabilistic classifier based on applying Bayes' theorem with strong (naive) independence assumptions. This classifier assumes that the presence or absence of a particular feature is unrelated to the presence or absence of any other feature, given the class variable. A naive Bayes classifier considers each of the features to contribute independently to the probability that a sample belongs to a given class, regardless of the presence or absence of the other features. Despite their naive design and apparently oversimplified assumptions, naive Bayes classifiers have worked quite well in many complex real-world situations. In fact, naive Bayes classifiers are simple, efficient and robust to noise and irrelevant attributes. However, they can only deal with symbolic data, although discretization techniques can be used to preprocess the data.

6.4.2.5   K-Nearest Neighbors, K-NN

K-Nearest neighbor [49] is a classification strategy that is an example of a "lazy learner". An object is classified by a majority vote of its neighbors, with the object being assigned to the class most common amongst its k nearest neighbors (where k is some user specified constant). If $k = 1$ (as it is the case in this thesis), then the object is simply assigned to the class of that single nearest neighbor. This method is more adequate for numerical data, although it can also deal with discrete values.

6.4.2.6   Multi-layer Perceptron, MLP

A multi-layer perceptron [50] is a feedforward artificial neural network model that maps sets of numerical input data onto a set of appropriate outputs. A MLP consists of multiple layers of nodes in a directed graph, with each layer fully connected to the next one. Except for the input nodes, each node is a neuron (or processing element) with a nonlinear activation function. MLP utilizes a supervised learning technique called back-propagation for training the network. MLP is a modification of the standard linear perceptron and can distinguish data that are not linearly separable.

6.4.2.7   AdaBoost, AB

AdaBoost ("Adaptive Boosting") [51], is a meta-algorithm which can be used in conjunction with many other learning algorithms to improve their performance (see a more detailed description, including pseudocode in Sect. 3.2.1). AdaBoost is adaptive in the sense that subsequent classifiers built are tweaked in favor of those instances misclassified by previous classifiers. It generates and calls a new weak classifier in each of a series of rounds. For each call, a distribution of weights is updated that indicates the importance of examples in the data set for the classification. On each round, the weights of each incorrectly classified example are increased, and the weights of each correctly classified example are decreased, so the new classifier focuses on the examples which have so far eluded correct classification. AdaBoost is sensitive to noisy data and outliers.

In order to evaluate the behavior of the feature selection methods after applying a classifier, several evaluation measures are usually employed, such as error, sensitivity, specificity, true positive rate, etc. (see Sect. 1.2 in Chap. 1).

6.5   Summary

In the previous chapters, we have discussed how to design an ensemble of methods for feature selection in a successful way, putting emphasis on issues such as how to choose the weak learners or how combine the partial results. But, eventually, the ensemble would need to be evaluated to see if it effectively works, and in this chapter we discussed important aspects on the evaluation of an ensemble. To start with, the methods conforming the ensemble must be diverse among them, but at the same time robust to different training data—i.e. stable. Finally, it is necessary to test if the selected features are the relevant ones, which can be made on artificial data provided that we know a priori the relevant features or using a classifier to evaluate the ultimate performance of the ensemble.

References

1. Brown, G., Wyatt, J.L., Tino, P.: Managing diversity in regression ensembles. J. Mach. Learn. **6**, 1621–1650 (2005)
2. Brown, G., Wyatt, J.L., Harris, R., Yao, X.: Diversity creation methods: a survey and categorisation. Inf. Fusion **6**(1), 5–20 (2005)
3. Lysiak, R., Kutzynski, M., Woloszynski, T.: Optimal selection of ensemble classifiers using measures of competence and diversity of base classifiers. Neurocomputing **126**, 29–35 (2014)
4. Visentini, I., Snidaro, L., Foresti, G.L.: Diversity-aware classifier ensemble selection via f-score. Inf. Fusion **28**, 24–43 (2016)
5. Dai, Q., Ye, R., Liu, Z.: Considering diversity and accuracy simultaneously for ensemble pruning. Appl. Soft Comput. **58**, 75–91 (2017)
6. Kuncheva, L.I., Whitaker, C.J.: Measures of diversity in classifier ensembles and their relationship with the ensemble accuracy. Mach. Learn. **51**(2), 181–207 (2003)
7. Kuncheva, L.I., Skurichinc, M., Duin, W.I.: An experimental study on diversity for bagging and boosting with linear classifiers. Inf. Fusion **3**, 245–258 (2002)
8. Kuncheva, L.I.: Special issue on diversity in multiple classifier systems. Inf. Fusion **6**(1), 1–116 (2005)
9. Cavalcanti, G.D.C., Oliveira, L.S., Moura, T.J.M., Carvalho, G.V.: Combining diversity measures for ensemble pruning. Pattern Recognit. Lett. **74**, 38–45 (2016)
10. Tsymbal, A., Pechenizkiy, M., Cunningham, P.: Diversity in search strategies for ensemble feature selection. Inf. Fusion **6**(1), 83–98 (2005)
11. Krawczyk, B., Woniak, M.: Diversity measures for one-class classifier ensembles. Neurocomputing **126**, 29–35 (2014)
12. Brodley, C.., Lane, T.: Creating and exploiting coverage and diversity. In: Proceedings of AAAI-96 Workshop on Integrating Multiple Learned Models, pp. 8–14 (1996)
13. Nogueira, S., Brown, G.: Measuring the stability of feature selection. In: Frasconi, P., Landwehr, N., Manco, G., Vreeken, J. (eds.) Machine Learning and Knowledge Discovery in Databases ECML PKDD 2016. Lecture Notes in Computer Science, vol. 9852. Springer, Berlin (2016)
14. Pes, B., Dess, N., Angioni, M.: Exploiting the ensemble paradigm for stable feature selection: a case study on high-dimensional genomic data. Inf. Fusion **35**, 132–147 (2017)
15. Seijo-Pardo, B., Porto-Díaz, I., Bolón-Canedo, V., Alonso-Betanzos, A.: Ensemble feature selection: homogeneous and heterogeneous approaches. Knowl. Based Syst. (2017). https://doi.org/10.1016/j.knosys.2016.11.017
16. Awada, W., Khoshgftaar, T.M., Dittman, D., Wald, R., Napolitano, A.: A review of the stability of feature selection techniques for bioinformatics data. In: Proceedings IEEE 13th International Conference on Information Reuse and Integration, pp. 356–363 (2012)
17. Altidor, W., Khoshgftaar, W., Van Hulse, J., Napolitano, A.: Ensemble feature ranking methods for data intensive computing applications. In: Furth, B., Escalante, A. (eds.) pp. 349–376. Spring, Berlin (2011)
18. Yang, F., Mao, K.Z.: Robust feature selection for microarray data based on multicriterion fusion. IEEE/ACM Trans. Comput. Biol. Bioinform. **8**(4), 1080–1092 (2011)
19. Ho, T.K.: The random subspace method for constructing decision forests. IEEE Trans. Pattern Anal. Mach. Intell. **20**(8), 832–844 (1998)
20. Lyerly, S.B.: The average Spearman rank correlation coefficient. Psychometrika **17**(4), 421–428 (1952)
21. Abdi, H.: The Kendall rank correlation coefficient. Encyclopedia of Measurement and Statistics, pp. 508–510. Sage, Thousand Oaks (2007)
22. Kolde, R., Laur, S., Adler, P., Vilo, J.: Robust rank aggregation for gene list integration and meta-analysis. Bioinformatics **28**(4), 573–580 (2012)
23. Willett, P.: Combination of similarity rankings using data fusion. J. Chem. Inf. Model. **53**(1), 1–10 (2013)

24. Joachims, T.: Optimizing search engines using clickthrough data. In: Proceedings of the eighth ACM SIGKDD international conference on Knowledge discovery and data mining, pp. 133–142 (2002)
25. Seijo-Pardo, B., Bolón-Canedo, V., Alonso-Betanzos, A.: Using a feature selection ensemble on DNA microarray datasets. In: Proceedings 24th European symposium on artificial neural networks, computational intelligence and machine learning (ESANN), pp 277–282 (2016)
26. Nogueira, S., Brown, G.: Measuring the stability of feature selection with applications to ensemble methods. In: Proceedings of International Workshop on Multiple Classifier Systems, pp. 135–146 (2015)
27. Abeel, T., Helleputte, T., Van de Peer, Y., Dupont, P., Saeys, Y.: Robust biomarker identification for cancer diagnosis with ensemble feature selection methods. Bioinformatics **26**(3), 392–398 (2009)
28. Ditzler, G., Polikar, R., Rosen, G.: A bootstrap based neyman-pearson test for identifying variable importance. IEEE Trans. Neural Netw. Learn. Syst. **26**(4), 880–886 (2015)
29. He, Z., Yu, W.: Stable feature selection for biomarker discovery. Comput. Biol. Chem. **34**(4), 215–225 (2010)
30. Saeys, Y., Abeel, T. and Van de Peer, Y., Robust feature selection using ensemble feature selection techniques, Machine learning and knowledge discovery in databases, ECML PKDD 2008, 34(4), 313–325, 2008
31. Kalousis, A., Prados, J., Hilario, M.: Stability of feature selection algorithms: a study on high-dimensional spaces. Knowl. Inf. Syst. **12**(1), 95–116 (2007)
32. Dunne, K., Cunningham, P., Azuaje, F.: Solutions to instability problems with sequential wrapper-based approaches to feature selection. J. Mach. Learn. Res. 1–22 (2002)
33. Kuncheva, L.I.: A stability index for feature selection. In: Proceedings of Artificial Intelligence and Applications, pp. 421–427 (2007)
34. Lustgarten, J.L., Gopalakrishnan, V., Visweswaran, S.: Measuring stability of feature selection in biomedical datasets. In: AMIA annual symposium proceedings, p. 406 (2009)
35. Wald, R., Khoshgoftaar, T.M. Napolitano, A.: Stability of filter-and wrapper-based feature subset selection. In: IEEE 25th International Conference on Tools with Artificial Intelligence (ICTAI), 2013, pp. 374–380 (2013)
36. Zhang, M., Zhang, L., Zou, J., Yao, C., Xiao, H., Liu, Q., Wang, J., Wang, D., Wang, Ch., Guo, Z.: Evaluating reproducibility of differential expression discoveries in microarray studies by considering correlated molecular changes. Bioinformatics **25**(13), 1662–1668 (2009)
37. Yu, L., Ding, C., Loscalzo, S.: Stable feature selection via dense feature groups. In: Proceedings of the 14th ACM SIGKDD international conference on knowledge discovery and data mining, pp. 803–811 (2008)
38. Somol, P., Novovicova, J.: Evaluating stability and comparing output of feature selectors that optimize feature subset cardinality. IEEE Trans. Pattern Anal. Mach. Intell. **32**(11), 1921–1939 (2010)
39. Křížek, P., Kittler, J., Hlaváč, V.: Improving stability of feature selection methods. In: Computer Analysis of Images and Patterns, pp. 929–936 (2007)
40. Voorhees, E.M.: Evaluation by highly relevant documents. In: Proceedings of the 24th annual international ACM SIGIR conference on research and development in information retrieval, pp. 613–622 (2001)
41. Jurman, G., Riccadonna, S., Visintainer, R., Furlanello, C.: Canberra distance on ranked lists. In: Proceedings of Advances in Ranking NIPS 09 Workshop, pp. 22–27 (2009)
42. Nogueira, S., Sechidis, K., Brown, G.: On the Use of Spearman's Rho to measure the stability of feature rankings. In: Iberian conference on pattern recognition and image analysis, pp. 381–391 (2017)
43. Belanche, L.A., González, F.F.: Review and evaluation of feature selection algorithms in synthetic problems (2017). http://arxiv.org/abs/1101.2320
44. Bolón-Canedo, V., Rego-Fernández, D., Peteiro-Barral, D., Alonso-Betanzos, A., Guijarro-Berdiñas, B., Sánchez-Maroño, N.: On the scalability of feature selection methods on high-dimensional data. Knowl. Inf. Syst. (2018)

45. Vapnik, V.N.: Statistical Learning Theory. Wiley, New Jersey (1998)
46. Fung, G., Mangasarian, O.L.: Proximal support vector machine classifiers. In: Proceedings of the seventh ACM SIGKDD international conference on knowledge discovery and data mining, pp. 77–86 (2001)
47. Quinlan, J.R., C4.5: Programs for Machine Learning. Morgan Kaufmann, Massachusetts (1993)
48. Rish, I.: An empirical study of the naive Bayes classifier. In: IJCAI 2001 workshop on empirical methods in artificial intelligence, pp. 41–46 (2001)
49. Aha, D.W., Kibler, D., Albert, M.K.: Instance-based learning algorithms. Mach. Learn. **6**(1), 37–66 (1991)
50. Hornik, K., Stinchcombe, M., White, H.: Multilayer feedforward networks are universal approximators. Neural Netw. **2**(5), 359–366 (1989)
51. Freund, Y., Schapire, R.E.: A decision-theoretic generalization of on-line learning and an application to boosting. In: Computational learning theory, pp. 23–37 (1995)

Chapter 7
Other Ensemble Approaches

Abstract This chapter describes several new fields, beside the feature selection pre-processing step (the theme of this book), in which ensembles have been successfully used. First, in Sect. 7.1, we introduce a very brief review of the different application fields in which ensembles have been applied, together with basic levels that are used to produce different ensemble designs, and a sample taxonomy. Then, in Sect. 7.2 basic ideas in ensemble classification design, one of the very first machine learning areas in which the idea of ensembles was applied, are stated. As there are many interesting and reference books in ensemble classification, we focus on describing the latest ideas in classification ensembles that address problems such as classification, stream data, missing data and imbalance data. Afterwards, the first attempts for applying the ensemble paradigm to the relatively new field of quantification are described in Sect. 7.3. In Sect. 7.4 we move to describing ensembles for clustering, another area in which ensembles have been increasingly popular. In Sect. 7.5 an attempt on ensembles for discretization is described and, finally, Sect. 7.6 summarizes and discusses the contents of this chapter.

This chapter is devoted to discuss briefly some of the fields, beside the feature selection one –the topic of this book– in which ensembles have contributed to improve performance. Machine learning and ensembles have been a good match over a large number of areas, such as classification, regression or clustering, and in a broad collection of applications. In this chapter, this "classical areas" of ensemble learning are briefly revisited, trying to pay more attention to the most recent approaches, in problems such as imbalance datasets, or in the presence of missing data. Moreover, some other new areas in which ensembles just have started to be applied (beside feature selection, the topic of this book), such as quantification or discretization are enumerated.

© Springer International Publishing AG, part of Springer Nature 2018   115
V. Bolón-Canedo and A. Alonso-Betanzos, *Recent Advances in Ensembles*
for Feature Selection, Intelligent Systems Reference Library 147,
https://doi.org/10.1007/978-3-319-90080-3_7

7.1   Introduction

Classification and regression were the first machine learning scenarios for which ensembles were devised. As it was mentioned in Sect. 3.1, the ensemble idea in supervised learning has been around in machine learning since the seminal works of Tuckey in 1977, in which two linear regression models were combined [1]; and of Dasarathy and Sheela [2], that suggested a partition of the input space using two or more classifiers. More than a decade after these initial works, the foundations for the well-known Adaboost [3, 4] were stated, demonstrating that by combining a number of the so-called weak classifiers (simple classifiers which classification performance is only slightly better than random classification), a strong classifier exhibiting better accuracy and stability than any of the individual methods, can be obtained. Since then, numerous methods have been proposed for ensembles of classifiers [5, 6], together with different measures of performance (for more details, please consult Chap. 6). There are several mechanisms that can be used to build ensembles of classifiers, which are detailed in Chap. 3. But as it occurs in many other fields of Machine Learning, there is not a clear winner method [7], and thus the field constitutes still an active area of research, and not only for classification, but also for regression, one-class, clustering, quantification, etc., as will be described in the sections below.

Ensembles have been applied to a vast variety of problems in real-world domains, such as medical diagnosis, recommender systems, sentiment analysis, text classification, spam detection, financial forecasting, weather forecasting, etc. [8, 9]. In 2015, a special issue on the topic " Hybrid and Ensemble Techniques: Recent Advances and Emerging Trends" [10] collected 22 articles in the field, and in January 2018, a search in Web of Science encountered 52 785 articles in the period 1990–2017, of which 21 876 articles (more than 40%) have been published in the last five years (2013–2017), which is a clear indicator of the activity in the field, that has contributions from many areas. Filtering only those results available in Web of Science Core Collection, the results obtained are shown in Fig. 7.1, in which it can be seen that although Computer Sciences and Engineering are the main areas of these publications, the ensemble paradigm is present in many other fields.

At present, there is a large number of ensemble techniques available, and thus several taxonomies of these methods have been proposed [9, 11, 12], so as to serve as a guide for the researchers and practitioners in the field. In Fig. 7.2 there are shown the different levels than can be used to construct different types of ensembles, that is, using different combination methods, using different base learners, using different feature subsets or using different subsets of the original dataset. As can be seen, these levels concentrate on the different parts of the ensemble than can be varied so as to obtain more accurate and robust learners. Based on these, the authors in [9] propose the taxonomy shown in Fig. 7.3, which divides the ensemble methods in two main groups: (1) those which are called "non-generative", that combine a set of existing base learners, and thus the emphasis lies on the selection and combination of the learner methods), and (2) the "generative" ensembles, that generate sets of base learners either acting on the base learner algorithm or in the structure of its input

Field: Research Areas	Record Count	% of **14551**	Bar Chart
COMPUTER SCIENCE	4015	18.964 %	▬
ENGINEERING	3460	16.342 %	▬
PHYSICS	2121	10.018 %	▪
CHEMISTRY	1620	7.652 %	▪
METEOROLOGY ATMOSPHERIC SCIENCES	1269	5.994 %	▪
SCIENCE TECHNOLOGY OTHER TOPICS	813	3.840 %	▪
BIOCHEMISTRY MOLECULAR BIOLOGY	710	3.353 %	▪
MATHEMATICS	698	3.297 %	▪
GEOLOGY	645	3.046 %	▪
WATER RESOURCES	633	2.990 %	▪

Fig. 7.1 Percentages of publications per research area of the Subject "Ensemble methods" using Web of Science for the period 2013–2017

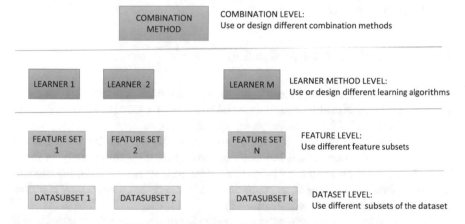

Fig. 7.2 Different levels that can be employed for ensemble design

with the aim of improving diversity and accuracy of the base learners. Thus, in this latter, the emphasis is on the construction of the diverse base learners, relegating to the background the combination technique.

Recently, emerging trends in the research field of ensemble techniques are those methods which are able to carry out online processing on data streams [13], supporting incremental learning [14–17], parallel learning for Big Data scenarios [18], or the problem of treating missing values [19, 20] or imbalanced data [21, 22]. Also, recently ensembles have been developed for one-class [23–25] or quantification problems [26, 27]. All these topics are very recent, as the publications are dated in the period 2016–2017, thus these appear to be the next hot topics for the field of ensemble learning for the following years.

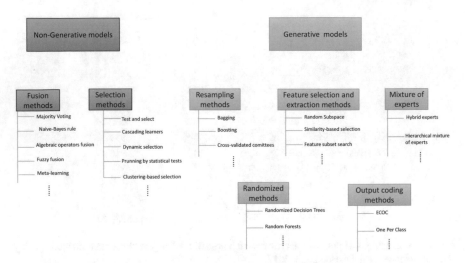

Fig. 7.3 Types of ensemble methods in a sample taxonomy. Some examples of specific types of ensemble methods are given

7.2   Ensembles for Classification

Classification is one of the pioneer fields of application for ensemble learning, and also perhaps the most prolific one, and several books and reviews have been published on the subject [6, 12, 28, 29]. However the application of ensembles for classification has been there from the late 70s, the interest of researchers in the topic has not decline, but on the contrary, it has been increasing their attention lately in many lines of classification related with dynamic selection, streaming data, imbalanced data, multi-label classification, concept drift, incremental learning, online learning, missing data or sentiment analysis [29–41].

As described in previous chapters, the main idea of the ensemble of classifiers is to design a committee of single (perhaps weak) classifiers, and combine their results so a classifier which can outperform the single ones can be obtained. Along the years, researchers have tried to understand and justify which is the reason why ensembles achieve better performance, further than the heuristic justification on the fact that more than one expert is better than a single one that all of us apply normally in our daily problems. Finally, there are several reason, of statistical, computational and representational background. Depending on the training algorithm and the training data used, the empirical estimate of the performance that will be obtained is a random value, and thus there is certain uncertainty associated with that estimation. For that reason, using a set of classifiers instead of a single one can be a better option, as the average of the outputs of the combination of classifiers might get a better generalization behavior. Computationally, we might take advantage of combining different classifiers that, individually obtain suboptimal solutions, in order to obtain better joint result. Making use of the divide and conquer principles and data fusion strategies are also another reasons [12].

As in the case of the feature selection ensembles, important aspects of classification ensembles are the ensemble cardinality (that is, the number of single classifiers that form the ensemble), as well as the diversity of those base classifiers, and not only in terms of predictive performance, but also in terms of memory and time consumption [29, 41–43]. Based on these evaluation measures, several taxonomies have been proposed both on ensemble types, combination methods and diversity generation methods [43, 44]. As it was said above, during the last years there are some areas of classification problems that have embraced ensembles as one attractive paradigm.

7.2.1 One-Class Classification

One of the interesting challenging problems in classification nowadays is one-class classification, that aims to learn when only data from one of the classes is available. To be more explicit, in a classical classification problem each unknown example is classified as belonging to one of all the available categories. However, there are different scenarios where the classification task consists in deciding whether a particular example fits a class, as it is the case in many real environments, such as fault detection in industrial machinery and robotics, intrusion detection in electronic security systems, video surveillance, etc. In order to handle appropriately these type of situations a one-class classification paradigm, in which one class (normal data or positive class) has to be distinguished from other classes (abnormal data), would be more appropriate. In these scenarios, the common situation is that the positive class is well represented in the training set, while the other classes are severely under-sampled or most commonly even nonexistent. The problem has received also other names, such as single classification, novelty detection, anomaly detection or outlier detection with subtle differences among them. The scarcity of abnormal examples might be due to several reasons, but the most frequent are their low frequency of occurrence or their extremely high costs. For example, in a machine monitoring system, measurements of the machine during its normal operational state are easy to obtain. However, measurements of failure are very expensive to collect as they would require a crash in the machine. Although one-class problems are quite frequent in real world, the truth is that there is a considerable lack of benchmark datasets for these problems, a fact that has slowed down progress, as researchers do not have a framework in which train and test their models. In consequence, the variety of learning models available for one-class problems is much smaller than in the standard classification area. One-class classification techniques can be grouped into three general categories: (a) Density estimation, (b) Reconstruction-based, and (c) Boundary-based techniques [45]. The first approach uses probabilistic methods that involve a density estimation of the target class, like mixture models and kernel density estimators. Reconstruction based methods involve training a regression model using the target class. These methods can autonomously model the underlying data, and when test data are presented to the system, the reconstruction error, defined to be the distance between the test vector and the output of the system, can be related to the novelty score. Finally, boundary methods try to model the boundary of the target class without focusing on the description of the underlying distribution.

Regarding the use of ensembles for one-class classification problems, in [46], the authors introduce several diversity measures applicable to the selection of one-class classifiers, aiming at introducing heterogeneity in the framework. For separating background in images an ensemble of local one-class classifiers is used in [23], and the same idea of local classifiers is devised in [47] combined with density analysis, and in [25], in which the authors split the target class into subsets that are used as input to one-class classifiers (SVM), that in turn are weighted to obtain a final multi-class classifiers. The method has the advantage of exhibiting a highly parallel structure of the solution process.

Nowadays, the advances in the ICT (Information and Communications Technology) field have contributed to the proliferation of big databases, usually distributed in several machines and in different locations. Performing predictive modeling, such as one-class classification, in this big data scenario is a difficult task, and as a consequence the majority of current one-class classification algorithms are unable to handle this new situation, or they do not scale properly, and thus distributed approaches have been developed, which in some cases use the same philosophy as in the case of the ensemble approaches, that is, applying a one-class classifier over a portion of the data, and then use a combination method that can give a final joint solution. For example, the work in [48] presented a framework for detecting anomalous behavior from terabytes of flight record data from distributed data sources that cannot be directly merged; in [49] the authors present a distributed version of the state of the art μ-SVM [50] algorithm, where several models are considered, each one determined using a given local data partition on a processor, and the goal is to find a global model. Other models based on convex-hull have been proposed, in which the geometrical structure of the convex hull (CH) is used to define the class boundary in one-class classification problems,as the one in [51], or the proposed in [24], that based on the previous one, makes a new proposal avoiding possible non-convex situations, and also approximates the n-dimensional convex-hull decision by means of random projections and an ensemble of convex-hull models in very low dimensions, which makes it suitable for larger dimensions in an acceptable execution time. The latter work has also an added interesting feature, it is also privacy preserving, that is, there is no data interchange among the different nodes.

7.2.2 Imbalanced Data

Many classification problems of the real world present imbalanced data, that is the number of samples of one or several of the classes of the problem is much lower than the other classes, in fact one-class scenarios might be considered as an extreme example of this type of situation. Standard classification methods present a problem in this situation, as they will have a bias towards the majority classes, while the minority classes are mostly ignored, as more general rules are preferred.

Ensemble techniques has been one of the paths followed by the researchers in order to confront the imbalance problem. An interesting review is done in [32], and as an added value the authors propose a taxonomy and a comparison of all the methods discussed. As ensemble methods are usually designed to boost accuracy, their direct application to datasets that are imbalanced is not worthwhile, and thus they are to be combined with other techniques that deal specifically with class imbalance. Re-sampling of imbalanced data is commonly used (over- or under-sampling) as it is independent of the classifiers being employed in the ensemble, and thus base classifiers do not need to be changed. Undersampling is a nonheuristic method that eliminates examples from the majority class, while oversampling replicates examples from the minority class. Their main drawbacks are increasing the probability of overfitting (for oversampling), and eliminating possible useful data (in the case of undersampling). The conclusion is that ensemble-based algorithms really make a difference, and although the complexity is increased by having more than a single classifier, they justify this increase by boosting the performance. Also, the authors recommend the use of simple approaches combining random undersampling techniques with bagging or boosting ensembles, as they exhibit a good balance between performance and complexity. Other works arrived to similar conclusions, as in [21, 52–57]. In the software tool KEEL [58] algorithms for treating imbalanced data, as well as other situations as missing data, streaming, etc., are available. See Chap. 9 for more details.

7.2.3 Data Streaming

In many real world problems, the classifiers need to learn dynamically, as their input is a stream of data, and thus the the target concept and its statistical properties might change over time, in a non predictable way. Learning classifiers from data streams is a relatively recent area in Machine Learning, that implies certain specific requirements in the methods used, being the most obvious ones a fast adaptation to change and low computation costs in both memory and time. There are however, other important challenges, as concept drift, feature drifts, novel classes, temporal dependencies, massive amount of data and/or features, limited amount of labelled instances, etc. [59]. In order to deal with data streams, most approaches adopt an incremental or on-line fashion, but also some adaptation mechanism should be devised in order to decide if the classifier should remain unchanged or not. For this reason, ensemble methods have been adopted as one of the most used solutions, as they can be integrated with drift detection mechanisms and incorporate dynamic updates, such as the selective removal (of the worst components of the current ensemble) or addition of classifiers (for example, building ensemble members on part of the input stream data). There are several works on the application of ensembles to data streams [13, 15–18], but in [59] the authors propose a taxonomy for these type of stream data classification ensembles, besides establishing some current and future trends in the field. In this taxonomy, besides the already known classification of the ensemble approaches using Combination and Diversity methods and Base Classifiers, a specific aspect of ensembles for data streams, named as "update dynamics" is introduced. This latter implies important peculiarities of those methods for stream learning, for example, strategies to cope with drifts, how learning is performed, and when to remove or add classifiers. Learning from data streams requires, beside accuracy, methods that should be efficient and able to adapt to changes in data. There are two main aspects to be taken into account: Cardinality and Learning mode. This introduces a new level in the taxonomy introduced in Fig. 7.3, that is shown in the expanded taxonomy in Fig. 7.4.

As said above, the two new aspects to be taken into account are Cardinality and Learning mode. Regarding cardinality, one needs to find a balance, because if too many classifiers are employed it is difficult to maintain diversity, and memory and time consumption may worsen considerably. In data streams ensemble classification, cardinality might be fixed a priori or dynamically. Intuitively, dynamic ensembles should have more adaptation power, but as they usually employ an heuristic in order to decide when to add or remove classifiers from the ensemble, their behavior might not be adequate to certain data streams, even in those cases in which a threshold that works as maximum number of classifiers is used [59–63]. Thus, as fixed strategies work reasonably well, besides avoiding the need of deciding an adequate heuristic, most methods work in this way. Examples of fixed cardinality are M^3 [64], MOOB [65] and BLAST [66], and of dynamic cardinality WOO [67] and SAE [68]. For a more complete list of data stream ensembles, please consult [59].

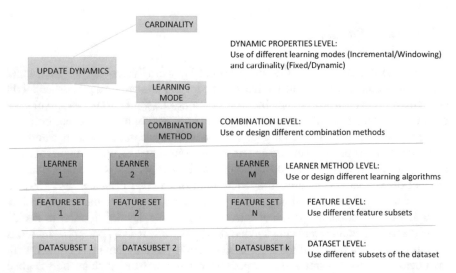

Fig. 7.4 Types of ensemble methods in a sample taxonomy that is expanded to include the peculiarities of updating dynamics that are needed in stream data ensembles

The second item, Learning Mode, is related with the stability-plasticity dilemma, that is the balance between the classifier learning new concepts and at the same time retaining knowledge learned previously. This balance is mandatory for any data stream classifier so to be able to adapt to concept drifts [69]. Ensemble-based algorithms, because of being formed by several models, can be more flexible with respect to concept drift adaptation. At this regard, ensembles can combine different types of incremental or window-based methods; retrain the models derived or either update them incrementally; using a proactive or a reactive adaptation, etc.[59, 69].

7.2.4 Missing Data

In real-world data it is also typical to found missing data, among other problems such as erroneous or corrupted data, noise or outliers, that need an answer from the machine learning side, so as to be able to deal with them. Missing data is a problem that nowadays might be very substantial in some datasets, for example in clinical studies that last several years, in which patients might not answer to certain questions in a questionnaire, or some new tests become available but were not at the beginning of the study, or a patient might just drop the study. In consequence, there are different mechanisms explaining the missingness of the data [70] and several mechanisms that can deal with them in Machine Learning [71]. The naive method of just deleting the instances with missing data is not possible, as many interesting patterns might be lost, and thus mechanisms that alleviate their effect while making use of the most data possible are needed.

Among others, ensemble methods have been one of the possible alternatives [72], that have the advantage of making it possible to include several different models that can deal with the different missing mechanisms. In [73] the authors describe an ensemble method that applies two different imputation methods (that is, substituting the missing values by estimations obtained by robust statistical methods or by machine learning methods, such as Bayesian, k-nearest neighbors, self-organizing maps or decision-tree models): a multiple one (the Bayesian multiple imputation), and a single one (the k-nearest neighbor), combining their results by voting. The same authors in [72] shown that by combining more methods (up to 7) considerably better performance results can be obtained. In [74], two concepts are used: multiple imputation and ensemble networks for devising two different ensembles, a univariate and a multivariate one. In both cases, the main idea is to use the uncertainty in the missing data, modelling it in terms of their probability distribution, to create different versions of the dataset, that are in turn employed to train different networks in an ensemble. Then, the missing values are filled in the training sets using those probability distributions, repeating the procedure several times, and thus obtaining different versions of the dataset, that is fed each one to a network model. Finally, the networks' outputs are averaged to produce the ensemble's final output.

The specific problem of dealing with missing values in time series is addressed in [75, 76], with the aim of providing an alternative for the missing value in signals coming from different sensors, for example, that will be used to estimate emotional states in an individual. In practical applications, data loss due to artifacts occurs frequently. In [75] the authors carried out a comparison using classifier fusion, that has shown significant increase in the accuracy in the recognition of emotional states. The authors tested two different approaches: ensembles using imputation (the missing features are imputed using median values) and ensembles using a feature-reduction approach (the features with missing values do not supply the ensemble with a classifier). Their conclusion is that the latter is comparable, and in some cases even better, than the ensemble with imputed values, an interesting fact for real-time approaches in which the complexity of the methods is an important restriction.

In [77], an ensemble of classifiers using random subspace selection as an alternative for dealing with missing values is proposed. The main differential aspect of this ensemble is that the missing values are not imputed, but the algorithm (named Learn++.MF) trains an ensemble of classifiers, each on a random subset of the available features. Instances with missing values are classified by the majority voting of those classifiers which training data did not include the missing features. In this way specific assumptions on the underlying data distribution are not needed, and the algorithm can deal with substantial amounts of missing data (30%) with only a slow gradual decline in performance as the amount of missing data increases.The algorithm assumes that the feature set is partially redundant, and that this redundancy is distributed randomly over the feature set. In [78] the idea is further explored, using the random subspace of input decimated ensembles, and the random subspace of support vector machines, in which classifiers are combined by the sum rule. Although their experimentation is restricted to medical datasets, they tried multiple imputation approaches based on random subspace, where each missing value is calculated considering a different cluster of the data. They have achieved a method that works well across several datasets, and which degradation is even lower than in the previous approach. Furthermore, their idea on clustering can be coupled with several missing imputation approaches allowing an improvement of the performance obtained by the standard imputation approaches alone.

Finally, the deep learning approaches have also entered the scene of ensembles for missing values in time series. In [19] an ensemble of multiple forecasting modules, based on a variant of the deep stacking network learning approach, is used. These modules are coupled employing dummy data, that is initially predicted using earlier points of the sequence of temporal data. These dummy data is progressively improved to best conform to the next parts of the sequence.

The different types of classifier ensembles above are still open research lines in which new developments are appearing constantly. Also, deep learning has made his way into the field of classification ensembles, with several new proposals in different areas of application [79, 80].

7.3 Ensembles for Quantification

As producing, saving, transferring, sharing, etc. data has become easier, inexpensive and rapid nowadays, and multiple sensors are available for measuring almost any activity that we can think of, "datification" of every process has transformed many sectors, as health or finance, into digital information and knowledge services. Thus, data has become available in large amounts, and among other tasks, Machine Learning can make use of it for a new task, that consists in producing aggregated estimations for a full sample rather than giving a specific prediction for each instance, as it is the task in classification. This new task is known as Quantification, a problem in which class prevalence $P(y)$ changes but $P(x|y)$ remains constant [81, 82], and its aim is to accurately estimate the number of cases belonging to each class (or class distribution) in a test set, using a training set that may have a substantially different distribution. When this shift occurs, the joint distribution of inputs and outputs changes between the training and testing phases.

As it was detailed in Chap. 6, diversity is a crucial aspect of ensembles' success. Diversity can be introduced into the ensemble by creating different training samples for each model, which is the basic idea of bagging. In that case, each model is trained with a data distribution that may be different from the original training set distribution. Following this idea, the possibility of developing ensemble versions of quantification algorithms is straightforward. Although at first sight it might appear that this is the same idea as in concept-drift ensembles for classification, it is not so. The first and main difference is that the concept does not change in quantification applications, while it does in concept-drift classification. For example, the concept of what is a positive opinion about a specific product does not change for a market analysis problem. The ensembles that deal with concept drift are usually designed to maintain a memory of models, that represent the evolving concept and thus past models that become valid again, are reused. In quantification tasks, however, this is impossible, as the concept does not change. The second important difference is that ensembles for concept drift are trained with successive samples. In quantification, the samples are generated according to the expected changes in the data distribution, with each one representing an specific and expected distribution change [26]. In this last referred work, the authors, who claim to have designed the first ensemble-based quantifier, propose to generate each training sample using the procedure shown in the left part (blue color) of Fig. 7.5 for binary quantification problems. First, the sample prevalence p_i is selected randomly in the [0,1] interval. Then, simple random sampling with replacement is performed with the positive class examples until obtaining their final number according to the chosen prevalence. The same operation is repeated for the negative class, which prevalence is $1 - p_i$. By changing the prevalence of each generated sample the desired diversity can be finally obtained. This procedure is repeated until the number of defined training samples is reached.

Three different baseline quantifiers were tested over 32 datasets, obtaining results that clearly outperform their single algorithm counterparts, and thus opening a new learning field for ensembles.

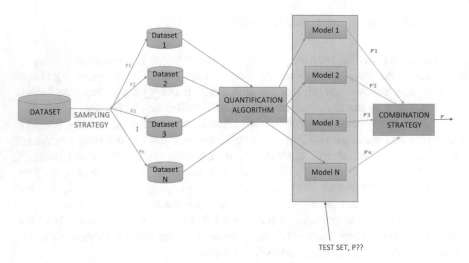

Fig. 7.5 An illustration of the quantification ensemble proposed in [26]. In the left side (in blue) the training generation process is depicted, while in the right hand side (in brown) the test part is depicted

7.4   Ensembles for Clustering

In recent years, the ensemble idea has been successfully used to tackle well known drawbacks of individual clustering algorithms [83]. The clustering problem, as well as seen before for other preprocessing techniques (as feature selection or discretization), appears to be another interesting field of application (as it is the case of the classical classification), for the use of combining multiple classifiers to solve difficult classification problems, using techniques such as bagging, boosting, etc as was detailed in Chap. 1. Analogously, cluster ensembles address the problem of combining multiple 'base clusterings' of the same set of objects into a single consolidated clustering, frequently called consensus solution.

There are however some aspects in clustering that complicate the problem and need to be taken into account, related with the fact that the optimal number of clusters in the consensus is not known in advance. Besides, the number of clusters that are obtained by each base clustering algorithm might very well be different. An additional complication is the fact that the base algorithms identify the groups using the original data, but they assign symbolic labels to each one. Base algorithms might then obtain different number of groups, and using different original data for each group and thus combining those labels across the different partial solutions is far from being simple. But beyond the expected improvement in accuracy provided by the averaging effect of many clustering algorithms aiming at the same goal, the potential motivations and benefits of using clustering ensembles are broader than those for using classification or regression ensembles, and some of those improvements, as quality and robustness, are similar to the ones obtained also for feature selection.

The raise in quality is due to the fact that the ensemble takes into account the biases of individual solutions [84], while clustering robustness is mainly due to the diversity of the employed methods, that allow to use approaches well suited for both low and high-dimensional metric spaces, and thus being able of providing adequate results on wider ranges of datasets. Specially interesting, for the topic of this book, is the research line followed by the work described in [85], in which feature diversity is produced using several different feature extraction techniques (that also aim at supervised model order selection) prior to the cluster ensemble, that employs the same model (k-means clustering using cosine distance). Some additional benefits for clustering ensembles are:

- Novelty, as the solution obtained by the ensemble is not reachable by single clustering methods.
- Stability and confidence estimation, providing clustering solutions with lower sensitivity to noise, outliers, or sampling variations. Besides, clustering uncertainty can be assessed from ensemble distributions [86].
- They constitute a new approach for model selection, by considering the match across the base solutions in order to provide the final number of clusters to use.
- If previous knowledge on several different possible groups of objects is available in the domain problem, the ensemble can help to integrate that information and obtain a more consolidated clustering. An example is the categorization of web pages using the document hierarchies available in several repositories together with the use of text analysis.
- Multiview clustering, parallelization and scalability. Analogously as in the other fields in which the ensemble idea is used, the ensemble can be constructed using the same data points but different clustering algorithms, or alternatively using the same clustering method over different partitions of the input data.

This last possibility enhances the final results, as often the objects to be clustered have different aspects that might provide for different clustering solutions.

In an example in Fig. 7.6, we can see three organizations that have three different base clusters over 6 different samples $\{X_1, X_2, X_3, X_4, X_5, X_6\}$ of a dataset.

In the case in which the objects to be clustered are distributed in origin, a cluster ensemble might be the appropriate solution to integrate them in a global, unique solution. In this last case, if all base clustering results for all objects are available in one place to perform the analysis, we rely on the basic idea. But many times, in real life applications this is not the case and then two types of distributed ensemble clustering are applicable:

- Column-distributed cluster ensemble, in which different base clustering results of the objects are at different locations. This is the case in which separate organizations have different base clusterings on the same set of objects, but the base clusterings cannot be shared among them for privacy concerns (as it is the case in the previous point, multiview clustering). However, if each organization has interest in a more robust consensus clustering, the cluster ensemble problem has to be solved in a column-distributed way. An example of this situation might be the case

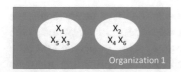

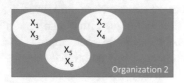

	K₁	K₂	K₃
X₁	1	1	1
X₂	2	2	3
X₃	1	1	2
X₄	2	2	3
X₅	1	3	2
X₆	2	3	4

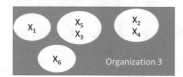

Fig. 7.6 An illustration of a clustering ensemble. Six different samples are clustered in three organizations using three different base clustering algorithms. Each column of the clustering matrix corresponds to each organization. Column-wise or row-wise clustering might be applied

Table 7.1 Column-wise consensus clustering

	k_1	k_2	.....	k_k
x_1	1	3	.....	5
x_2	1	1	.....	3
.	.	.	.....	.
.	.	.	.....	.
x_n	5	5	.....	1

in which different e-commerce vendors have different customer segmentations on the same customer base [87]. This type is also called "feature distributed clustering" (FDC), as different base clusterings are built by selecting different subsets of the features but using all the data points.

In Table 7.1 it can be seen a clustering matrix M that is a set of k columns, that contain k base clustering results $\{k_j, j_1^k\}$ for a dataset of n samples. As the aim is to derive a consensus clustering, we need to find out the correspondence between the different base clusters generated by the different algorithms. Thus, the cluster correspondence problem is hard to solve efficiently, increasing the complexity even more in the case that the different clustering algorithms generate different numbers of clusters.

- Row-distributed cluster ensembles (see an example in Table 7.2), in which different objects (rows) are at different locations, as for example might occur when different

Table 7.2 Row-wise consensus clustering

	k_1	k_2	.....	k_k
x_1	1	3	.....	5
x_2	1	1	.....	3
.	.	.	.....	.
.	.	.	.....	.
x_n	5	5	.....	1

subsets of the original dataset are owned by different organizations, or cannot be put together in one place due to size, communication, or privacy constraints. Of course, there is the possibility of applying a distributed clustering algorithm, but if there are restrictions on sharing or privacy, the results of the different subsets cannot be transmitted either to a central node for analysis. However, combining the results on different subsets helps to generate a more reasonable ensemble clustering, and thus a consensus clustering should be learned in a row-distributed manner. The clustering matrix M is now a set of N rows of k-dimensional feature vectors $\{x_i, i_1^n\}$, and the cluster ensemble aims at finding a clustering consensus for the feature vectors.

Another name of this type of ensemble clustering is "Object Distributed Clustering" as the base clusters are built using different subsets of the data points, but employing all the features. Again, for a customer segmentation use case, different companies might have different subsets of customers, and although a base clustering on all the customers dataset can be performed using privacy preserving clustering algorithms, the cluster assignments of the customer subsets for each vendor is private information, and the companies might not want to share it directly for the purposes of forming a consensus clustering. Thus, it will be desirable to have cluster ensemble algorithms handle such "row-distributed" base clusterings.

In summary, cluster ensembles combine the results obtained by different clustering algorithms through a consensus function, in order to obtain a more robust, stable and accurate solution. Hence, two fundamental components of the ensemble are (1) the mechanism that is used to generate the initial partitions, that should generate the necessary diversity, and (2) the consensus function used to combine these partitions into a final result. Regarding diversity several methods can be used, such as employing different clustering algorithms over the same or different dataset; using the same clustering algorithm but with different initialization, parameter values or built-in randomness, or using different partitions of the dataset; data resampling, etc. Finding an adequate consensus function is however a hard task, that will be discussed below in Sect. 7.4.1.

7.4.1 Types of Clustering Ensembles

There are several clustering ensemble algorithms regarding the consensus function types used. The consensus functions are employed to combine the different individual clusters obtained, aiming at ensuring a symmetrical and unbiased general agreement with respect to all the component partitions, and solving also the label problem, as each contributing partition has their own partial labels. As patterns are unlabelled there is no explicit correspondence between the labels obtained by the different partitions. Combining the multiple clusters can be also viewed as a median partition with respect to the given partitions, which is proven to be an NP-complete problem. The most well-known consensus functions can be classified in one of the types below:

- Relabeling, also called voting or direct approach. The basic idea is to have a reference partition, that can be one of the ensemble or a different one, and then relabel all partitions according to the reference. This is achieved by permuting the cluster labels such that the best agreement between the labels of the two partitions being compared is obtained. It is assumed that the number of clusters in every partition of the ensemble is the same as in the target partition, and that the number of clusters (k) of this target partition is known. The complexity of these methods is $k!$, a number that can be reduced to $\mathcal{O}(k^3)$ if it is employed the Hungarian method for the minimal weight bipartite matching problem [88]. Some well-known methods of this type are [89, 90].
- Graph-based models are the most popular, they work by converting the results of the base clusterings to a hypergraph or a graph, to which later graph partitioning algorithms are applied and, as a result, ensemble clusters are obtained. The problem of consensus clustering is reduced to finding the minimum-cut of the hypergraph into k components. Some of the most well-known algorithms of this type are the cluster-based similarity partitioning algorithm (CSPA) [91], the weighted bipartite partitioning algorithm (WBPA) [92], the Weighted Spectral Cluster Ensemble (WSCE)[93], the two last being able to deal with high-dimensional datasets. Hypergraph partitioning is also an NP-hard problem, but there are efficient heuristics to solve the k (the number of clusters) way min-cut partitioning problem, some of them with complexity $\mathcal{O}(|\varepsilon|)$ (being ε the number of hyperedges of the graph).
- Matrix-based models, which main idea is to convert the base clustering matrix into another matrix such as co-association matrix [94], consensus matrix [95] or nonnegative matrix [96], and then use matrix operations to get the results of the cluster ensemble. The main drawback of these methods is that their computational complexity is very high.
- Probabilistic models, in which the algorithms make use of statistic properties of base clustering results in order to achieve a consensus clustering. The most common models use either the Mutual Information or Finite mixture models. In the case of the Finite mixture models, labels are modeled as random variables drawn from a probability distribution described as a mixture of multinomial component densities. The aim of the consensus clustering is formulated as a maximum likelihood estimation problem, using the Expectation Maximization Algorithm (EM)

[97] to solve it [86]. Another set of approaches formulate the objective function of a clustering ensemble as the mutual information between the empirical probabilistic distribution of labels in the consensus partition and in the ensemble, or use other information theory based approaches [98, 99]. Finally, in [87] the authors propose a Bayesian approach, a mixed-membership model for learning cluster ensembles, that has the added advantage of avoiding completely the cluster label correspondence problems appearing in the graph based approaches, and besides can deal with situations of missing values and with both row and column-distributed cluster ensembles.

- Recently, some other authors [98, 100–103] have used approaches based on evolutionary algorithms, employing the searching capability of genetic algorithms to derive a consensus clustering from clustering ensembles. They propose a new consensus function based on genetic algorithms to find an almost median partition. The clustering metric used by this function is the sum of the entropy-based dissimilarity of the consensus clustering from the component clusterings in the ensemble. In [103] new consensus functions using genetic algorithms and three different fitness scores (normalized mutual information (NMI), adjusted mutual information (AMI) and generalized conditional entropy (GCE)) are devised and tested over different scenarios of row and column distributed clustering ensembles with good results in both stability and accuracy.

Concerning the evaluation measures for clustering ensembles, again those described in Chap. 6 are of application here, although some new specific measures were devised in [104–107].

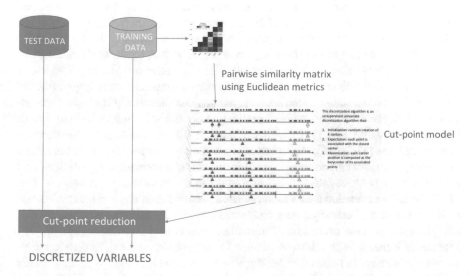

Fig. 7.7 A scheme of the discretization method based on clustering ensembles

7.5 Ensembles for Other Preprocessing Steps: Discretization

In many cases, data appears in the form of continuous values. If besides their number is huge, building models for such data might be very difficult. Moreover, there are many data mining algorithms (feature selection methods based on mutual information, for example, or decision trees, among the most popular), that operate only over discrete variables. Discretization is a preprocessing technique [108] that aims at transforming continuous functions or variables into nominal (discrete) counterparts, reducing the number of values a continuous variable has by grouping them into a number, b, of intervals or bins. Thus, discretization is oftentimes regarded as a data reduction mechanism, producing typically disjoint intervals that mutually cover the continuous value range of the attribute. As said before, discretization might be unavoidable if the model to be used afterwards is not able to handle continuous variables. However, although the subsequent algorithm is capable of working with continuous variables, discretization is applied so as to improve both speed and accuracy, nevertheless at the cost of assuming the loss of information that comes with the reducing dimensionality. In general, discretization constitutes an important preprocessing step that can greatly influence the performance of a machine learning system, and thus numerous discretization methods have been developed. Discretization can be performed in a supervised, that is using class information, or unsupervised way, not using class information. If class labels are available in the training dataset, then discretization methods aim at maximizing the interdependence between the variable values and the output class labels, and thus also minimizing information loss in the transformation from continuous to discrete values. Among the supervised discretization approaches, Fayyad and Irani's entropy-based discretization algorithm [109] is the most commonly used, probably due to its comprehensibility and quite good performance. Regarding the unsupervised approaches, equal-width binning and equal-frequency binning are the most popular. Discretization can also be univariate or multivariate. In the univariate algorithms, only one continuous feature is quantified at a time while multivariate discretization considers simultaneously multiple features. Two other decisions related with discretization are the selection of the number of bins and their width. For a taxonomy on discretization methods, and basic descriptions please see [110, 111]. Although univariate discretization algorithms are most commonly used, one important drawback is that useful information regarding natural groups, hidden patterns and correlation among the attributes will be probably lost. In [112] the authors develop a new unsupervised discretization method that makes use of the novel idea of encoding data clusters and similarity using a euclidean metric, into the discretization procedure. The method uses cluster ensembles (by applying repeatedly k-means with different values of k to the training data), to discover similarities between data points that belong to adjacent intervals, and when that occurs the cut-point is dropped. Thus, a new pruning method that exploits natural groups as an explicit constraint to the traditional cut-point determination techniques used in classical discretization algorithms is devised, as it is shown in Fig. 7.7.

7.6 Summary

In all the chapters of this book, we have discussed how to design and evaluate an ensemble of methods for feature selection. In this chapter, we made a review of other areas in which ensemble learning has been successful. First, we visit the ensemble ideas in classification and regression, the pioneer fields in ensemble learning. As there are many good books devoted to the description of classification ensembles, we just summarized the basic ideas, and try to concisely enumerate the new problems in which ensembles are being applied as now, as it is the case of imbalance datasets, missing data, data streams, one-class classification, etc. Next, clustering ensembles, another successful line of applications, are also shortly described. Finally, areas in which ensembles have just loomed up, as quantification or discretization, are reviewed with the aim of boosting researcher for more proposals in these and other fields of machine learning.

References

1. Tukey, J.W.: Exploratory Data Analysis. Addison-Wesley, Reading (1977)
2. Dasarathy, B.V., Sheela, B.V.: Composite classifier system design: concepts and methodology. Proc. IEEE **67**(5), 708–713 (1979)
3. Freund, Y.: Boosting a weak learning algorithm by majority. Inf. Comput. **121**(82), 256–285 (1995)
4. Freund, Y., Schapire, R.E.: Experiments with a new boosting algorithm. In: Machine Learning: Proceedings of the Thirteenth International Conference, pp. 325–332 (1996)
5. Dietterich, T.G.: An experimental comparison of three methods for constructing ensembles of decision trees: bagging, boosting, and randomization. Mach. Learn. **40**, 139–157 (2000)
6. Rokach, L.: Ensemble-based classifiers. Artif. Intell. Rev. **33**(1), 1–39 (2010)
7. Villada, R., Drissi, Y.: A perspective view and survey of meta-learning. Artif. Intell. Rev. **18**, 77–95 (2002)
8. Okun, O.: Applications of Supervised and Unsupervised Ensemble Methods. Springer, Berlin (2009)
9. Re, M., Valentini, G.: Ensemble Methods: A Review In Advances in Machine Learning and Data Mining for Astronomy, pp. 563–594. Chapman & Hall, Boca Raton (2012)
10. Kazienko, P., Lughofer, E., Trawinski , B.: Special issue on Hybrid and ensemble techniques: recent advances and emerging trends. Soft Comput. **19**(12), 3353–3355 (2015)
11. Sharkey, A.J.C.: Types of multinet systems. In: Roli, F., Kittler, J. (eds.) Proceedings of Multiple Classifier Systems. MCS 2002. Lecture Notes in Computer Science, vol. 2364, pp. 108–117. Springer, Berlin (2002)
12. Kuncheva, L.I.: Combining Pattern Classifiers: Methods and Algorithms, 2nd edn. Wiley, New York (2014)
13. Gama, J.: Knowledge Discovery from Data Streams. Chapman & Hall/CRC, Boca Raton (2010)
14. Hu, J., Li, T.R., Luo, C., Fujita, H., Yang, Y.: Incremental fuzzy cluster ensemble learning based on rough set theory. Knowl.-Based Syst. **132**, 144–155 (2017)
15. Duan, F., Dai, L.: Recognizing the gradual changes in sEMG characteristics based on incremental learning of wavelet neural network ensemble. IEEE Trans. Industr. Electron. **64**(5), 4276–4286 (2017)

16. Khan, I., Huang, J.Z., Ivanov, K.: Incremental density-based ensemble clustering over evolving data streams. Neurocomputing **191**, 34–43 (2016)
17. Yu, Z.W., Luo, P.N., You, J.N., Wong, H.S., Leung, H., Wu, S., Zhang, J., Han, G.Q.: Incremental Semi-Supervised Clustering Ensemble for High Dimensional Data Clustering. IEEE Trans. Knowl. Data Eng. **28**(3), 701–714 (2016)
18. Huang, S., Wang, B.T., Qiu, J.H., Yao, J.T., Wang, G.R., Yu, G.: Parallel ensemble of online sequential extreme learning machine based on MapReduce. Neurocomputing **174**, 352–367 (2016)
19. Das, M., Ghosh, S.K.: A deep-learning-based forecasting ensemble to predict missing data for remote sensing analysis. IEEE J. Sel. Top. Appl. Earth Obs. Remote Sens. **10**(12), 5228–5236 (2017)
20. Gao, H., Jian, S.L., Peng, Y.X., Liu, X.W.: A subspace ensemble framework for classification with high dimensional missing data. Multidimens. Syst. Signal Process. **28**(4), 1309–1324 (2017)
21. Lu, W., Li, Z., Chu, J.: Adaptive ensemble undersampling-boost: a novel learning framework for imbalanced data. J. Syst. Softw. **132**, 272–282 (2017)
22. Lin, W.C., Tsai, C.F., Hu, Y.H., Jhang, J.S.: Clustering-based undersampling in class-imbalanced data. Inf. Sci. **409**, 17–26 (2017)
23. Silva, C., Bouwmans, T., Frelicot, C.: Superpixel-based online wagging one-class ensemble for feature selection in foreground/background separation. Pattern Recogn. Lett. **100**, 144–151 (2017)
24. Fernández-Francos, D., Fontenla-Romero, O., Alonso-Betanzos, A.: One-class convex hull-based algorithm for classification in distributed environments. IEEE Trans. Syst. Man Cybern. Syst. (2017). https://doi.org/10.1109/TSMC.2017.2771341
25. Krawczyk, B., Cyganek, B.: Selecting locally specialised classifiers for one-class classification ensembles. Pattern Anal. Appl. **20**(2), 427–439 (2017)
26. Pérez-Gallego, P.J., Quevedo-Pérez, J.R., Coz-Velasco, J.J.: Using ensembles for problems with characterizable changes in data distribution: a case study on quantification. Inf. Fusion **34**, 87–100 (2017). https://doi.org/10.1016/j.inffus.2016.07.001
27. Mallet, V., Herlin, I.: Quantification of uncertainties from ensembles of simulations. In: International Meeting Foreknowledge Assessment Series (2016). http://www.foreknowledge2016.com/
28. Brown, G.: Ensemble learning. In: Sammut, C., Webb, G.I. (eds.) Encyclopedia of Machine Learning. Springer, Berlin (2010)
29. Pietruczuk, L., Rutkowski, L., Jaworski, M., Duda, P.: How to adjust an ensemble size in stream data mining. Inf. Sci. **381**, 46–54 (2017)
30. Yin, Z., Zhao, M.Y., Wang, Y.X., Yang, J.D., Zhang, J.H.: Recognition of emotions using multimodal physiological signals and an ensemble deep learning model. Comput. Methods Programs Biomed. **140**, 93–110 (2017)
31. Wozniak, M., Grana, M., Corchado, E.: A survey of multiple classifier systems as hybrid systems. Inf. Fusion **16**, 3–17 (2014)
32. Galar, M., Fernández, A., Barrenechea, E., Bustince, H., Herrera, F.: A review on ensembles for the class imbalance problem: bagging-, boosting-, and hybrid-based approaches. IEEE Trans. Syst. Man Cybern. Part C- Appl. Rev. **42**(4), 463–484 (2012)
33. Elwell, R., Polikar, R.: Incremental learning of concept drift in nonstationary environments. IEEE Trans. Neural Netw. **22**(10), 1517–1531 (2011)
34. Krawczyk, B., Minku, L.L., Gama, J., Stefanowski, J., Wozmiak, M.: Ensemble learning for data stream analysis. A survey. Inf. Fusion **37**, 132–156 (2017)
35. Cruz, R.M.O., Sabourin, R., Cavalcanti, G.D.: Dynamic classifier selection: recent advances and perspectives. Inf. Fusion **41**, 195–216 (2018)
36. Brun, A.L., Britto Jr., A.S., Oliveira, L.S., Enembreak, F., Sabourin, F.: A framework for dynamic classifier selection oriented by the classification problem difficulty. Pattern Recogn. **76**, 175–190 (2018)

37. Armano, G., Tamponi, E.: Building forests of local trees. Pattern Recognit. **76**, 380–390 (2018)
38. Mayano, J.M., Gibaja, E.L., Cios, K.J., Ventura, S.: Review of ensembles of multi-label classifiers: models, experimental study and prospects. Inf. Fusion **44**, 33–45 (2018)
39. Monidipa, D., Ghosh, S.K.: A deep-learning-based forecasting ensemble to predict missing data for remote sensing analysis. IEEE J. Sel. Top. Appl. Earth Obs. Remote Sens. **10**(12), 5228–5236 (2017)
40. Yan, Y.T., Zhang, Y.P., Zhang, Y.W., Du, X.Q.: A selective neural network ensemble classification for incomplete data. Int. J. Mach. Learn. Cybernet. **8**(5), 1513–1524 (2017)
41. Bonab, H.R., Fazli, C.: Less is more: a comprehensive framework for the number of components of ensemble classifiers. IEEE Trans. Neural Netw. Learn. Syst
42. Hernández-Lobato, D., Martínez-Muñoz, G., Suárez, A.: How large should ensembles of classifiers be? Pattern Recogn. **47**(5), 1323–1336 (2017)
43. Brown, G., Wyatt, J., Harris, R., Yao, X.: Diversity creation methods: a survey and categorization. Inf. Fusion **6**(1), 5–20 (2005)
44. Tsoumakas, G., Partalas, I., Vlahavas, I.: A taxonomy and short review of ensemble selection. In: ECAI 08, Workshop on Supervised and Unsupervised Ensemble Methods and Their Applications (2008)
45. Khan, S.S., Madden, M.G.: One-class classification: taxonomy of study and review of techniques. Knowl. Eng. Rev. **29**(3), 345–374 (2014)
46. Dib, G., Karpenko, O., Koricho, E., Khomenko, A., Haq, M., Udpa, L.: Ensembles of novelty detection classifiers for structural health monitoring using guided waves. Smart Mater. Struct. **27**(1) (2018). https://doi.org/10.1088/1361-665X/aa973f
47. Liu, J., Miao, Q., Sun, Y., Song, J., Quan, Y.: Modular ensembles for one-class classification based on density analysis. Neurocomputing **171**, 262–276 (2016)
48. Zhou, X., Zhong, Y., Cai, L.: Anomaly detection from distributed flight record data for aircraft health management. In: Proceedings of International Conference on Computational and Information Sciences, pp 156–159 (2010)
49. Castillo, E., Peteiro-Barral, D., Guijarro-Berdiñas, B., Fontenla-Romero, O.: Distributed one-class support vector machine. Int. J. Neural Syst. **25**(7), 1550029 (2015)
50. Schölkopf, B., Williamson, R.C., Smola, A.J., Shawe-Taylor, J., Platt, J.C.: Support vector method for novelty detection. In: Advances in Neural Information Processing Systems, NIPS '00, pp. 582–588 (2000)
51. Casale, P., Pujol, O., Radeva, P.: Approximate polytope ensemble for one-class classification. Pattern Recogn. **47**(2), 854–864 (2014)
52. Galar, M., Fernández, A., Barrenechea, E., Herrera, F.: EUSBoost: enhancing ensembles for highly imbalanced data-sets by evolutionary undersampling. Pattern Recogn. **46**(12), 3460–3471 (2013)
53. Salunkhe, U.R., Suresh, N.M.: Classifier ensemble design for imbalanced data classification: a hybrid approach. Procedia Comput. Sci. **85**, 725–732 (2016)
54. Wang, Q., Luo, Z., Huang, J.C., Feng, Y.H., Liu, Z.: A novel ensemble method for imbalanced data learning: bagging of extrapolation-SMOTE SVM. Comput. Intell. Neurosci. pp. 1827016 (2017). https://doi.org/10.1155/2017/1827016
55. Sun, Y., Kamel, M., Wong, A., Wang, Y.: Cost-sensitive boosting for classification of imbalanced data. Pattern Recogn. **40**, 3358–3378 (2007)
56. Blaszczynski, J., Deckert, M., Stefanowski, J., Wilk, S.: Integrating selective pre-processing of imbalanced data with ivotes ensemble. In: 7th International Conference on Rough Sets and Current Trends in Computing (RSCTC2010), LNCS 6086, pp. 148–157. Springer (2010)
57. Wang, S., Yao, X.: Diversity analysis on imbalanced data sets by using ensemble models. In: IEEE Symposium Series on Computational Intelligence and Data Mining (IEEE CIDM 2009), pp. 324–331 (2009)
58. Triguero, I., González, S., Moyano, J.M., García, S., Alcalá-Fernández, J., Luengo, J., Fernández, A., del Jesus, M.J., Sánchez, L., Herrera, F.: KEEL 3.0: an open source software for multi-stage analysis in data mining. Int. J. Comput. Intell. Syst. **10**, 1238–1249 (2017)

59. Gomes, H.M., Barddal, J.P., Enembreck, F., Bifet, A.: A survey on ensemble learning for data stream classification. ACM Comput. Surv. **50**(2), 1–23 (2017)
60. Barddal, J.P., Gomes, H.M., Enembreck, F., Pfahringer, B.: A survey on feature drift adaptation: definition, benchmark, challenges and future directions. J. Syst. Softw. **127**, 278–294 (2017)
61. Bifet, A., Holmes, G., Pfahringer, B.: Leveraging bagging for evolving data streams. In: Proceeding ECML PKDD'10, European Conference on Machine Learning and Knowledge Discovery in Databases: Part I, pp. 135-150 (2010)
62. Brzezinski, D., Stefanowski, J.: Combining block-based and online methods in learning ensembles from concept drifting data streams. Inf. Sci. **265**, 50–67 (2014)
63. Brzezinski, D., Stefanowski, J.: Ensemble diversity in evolving data streams. In: Proceedings of the International Conference on Discovery Science, pp. 229–244. Springer (2016)
64. Parker, B.S., Khan, L., Bifet, A.: Incremental ensemble classifier addressing nonstationary fast data streams. In: Proceedings of the 2014 IEEE International Conference on Data Mining Workshop (ICDMW), pp 716–723. IEEE (2014)
65. Wang, S., Minku, L.L., Yao, X.: Resampling-based ensemble methods for online class imbalance learning. IEEE Trans. Knowl. Data Eng. **27**(5), 1356–1368 (2015)
66. van Rijn, J.N., Holmes, G., Pfahringer, B., Vanschoren, J.: Having a blast: metalearning and heterogeneous ensembles for data streams. In: Proceedings of the 2015 IEEE International Conference on Data Mining (ICDM), pp 1003–1008. IEEE (2015)
67. Ryu, J.W., Kantardzic, M.M., Kim, M.W.: Efficiently maintaining the performance of an ensemble classifier in streaming data. In: Convergence and Hybrid Information Technology, pp. 533–540. Springer (2012)
68. Gomes, H.M., Enembreck, F.: SAE: social adaptive ensemble classifier for data streams. In: Proceedings of the 2013 IEEE Symposium on Computational Intelligence and Data Mining (CIDM), pp 199–206 (2013). https://doi.org/10.1109/CIDM.2013.6597237
69. Gama, J., Žliobaitė, I., Bifet, A., Pechenizkiy, M., Bouchachia, A.: A survey on concept drift adaptation. ACM Comput. Surv. **46**(4) (2014)
70. Schafer, J.L., Graham, J.W.: Missing data: our view of the state of the art. Psychol. Methods **7**(2), 147–177 (2002)
71. Saar-Tsechansky, M., Provost, F.: Handling missing values when applying classification models. J. Mach. Learn. Res **8**, 1623–1657 (2007)
72. Twala, B., Cartwright, M.: Ensemble missing data techniques for software effort prediction. Intell. Data Anal. **14**, 299–331 (2010)
73. Twala, B., Cartwright, M.: Ensemble imputation methods for missing software engineering data. In: Proceedings of 11th IEEE Int. Software metric Symposium (2005)
74. Hassan, M.M., Atiya, A.F., El Gayar, N., El-Fouly, R.: Novel ensemble techniques for regression with missing data. New Math. Nat. Comput. **5** (2009)
75. Setz, C., Schumm, J., Lorenz, C., Arnrich, B., Tröster, G.: Using ensemble classifier systems for handling missing data in emotion recognition from physiology: one step towards a practical system. In: Proceedings of 3rd International Conference on Affective Computing and Intelligent Interaction and Workshops, pp. 1–8 (2009)
76. Moahmed, T.A., El Gayar, N., Atiya, A.F.: Forward and backward forecasting ensembles for the estimation of time series missing data. In: IAPR Workshop on ANN in Pattern Recognition, Lecture Notes in Computer Science, vol. 8774, pp. 93–104 (2014)
77. Polikar, R., DePasquale, J., Mohammed, H.S., Brown, G., Kuncheva, L.I.: Learn++.MF: A random subspace approach for the missing feature problem. Pattern Recogn. **43**, 3817–3832 (2010)
78. Nanni, L., Lumini, A., Brahnam, S.: A classifier ensemble approach for the missing value problem. Artif. Intell. Med. **55**(1), 37–50 (2012)
79. Rad, N.M., Kia, S.M., Zarbo, C., van Laarhoven, T., Jurman, G., Venuti, P., Marchiori, E., Furlanello, C.: Deep learning for automatic stereotypical motor movement detection using wearable sensors in autism spectrum disorders. Sig. Process. **144**, 180–191 (2018)

80. Xiao, Y.W., Wu, J., Lin, Z.L., Zhao, X.D.: A deep learning-based multi-model ensemble method for cancer prediction. Comput. Methods Programs Biomed. **153**, 1–9 (2018)
81. Forman, G.: Quantifying counts and costs via classification. Data Min. Knowl. Discov. **17**, 164–206 (2008)
82. Barranquero, J., Díez, J., Del Coz, J.J.: Quantification-oriented learning based on reliable classifiers. Pattern Recogn. **48**(2), 591–604 (2015)
83. Ghosh, J., Acharya, A.: Cluster ensembles. WiREs Data Min. Knowl. Discov. **1**(4), 305–315 (2011)
84. Kuncheva, L.I., Hadjitodorov, S.T.: Using diversity in cluster ensemble. Proc. IEEE Int. Conf. Syst. Man Cybern. **2**, 1214–1219 (2004)
85. Sevillano, X., Cobo, G., Alías, F., Socoró, J.C.: Feature diversity in cluster ensembles for robust document clustering. In: SIGIR '06 Proceedings of the 29th Annual International ACM SIGIR Conference on Research and Development in Information Retrieval, pp. 697–698 (2006)
86. Topchy, A.P., Jain, A.K., Punch, W.F.: Clustering ensembles: models of consensus and weak partitions. IEEE Trans. Pattern Anal. Mach. Intell. **27**(12), 1866–1881 (2005)
87. Wang, H., Shan, H., Banerjee, A.: Bayesian cluster ensembles. J. Stat. Anal. Data Min. **4**(1), 54–70 (2011)
88. Kuhn, H.W.: The Hungarian method for the assignment problem. Naval Res. Logic. Quart. **2**, 83–97 (1955)
89. Dudoit, S., Fridiyand, J.: Bagging to improve the accuracy of a clustering procedure. Bioinformatics **19**(9), 1090–1099 (2003)
90. Hong, Y., Kwong, S., Chang, Y., Ren, Q.: Unsupervised feature selection using clustering ensembles and population based incremental learning algorithm. Pattern Recogn. **41**(9), 2742–2756 (2008)
91. Streh, A., Ghosh, J.: Cluster ensembles - a knowledge reuse framework for combining multiple partitions. J. Mach. Learn. Res. **3**, 583–617 (2002)
92. Domeniconi, C., Al-Razgan, M.: Weighted cluster ensembles: methods and analysis. ACM Trans. Knowl. Discov. Data **2**(4), 1–40 (2009)
93. Yousefnezhad, M., Zhang, D.: Weighted spectral cluster ensemble. In: Proceedings of IEEE International Conference on Data Mining 2015, pp. 549–558 (2015)
94. Fred, A.L.N., Jain, A.K.: Data clustering using evidence accumulation. In: Proceedings of 16th International Conference on Pattern Recognition-ICPR, pp. 276–280 (2002)
95. Monti, S., Tamayo, P., Mesirov, J., Golub, T.: Consensus clustering: a resampling-based method for class discovery and visualization of gene expression microarray data. Mach. Learn. J. **52**, 91–118 (2003)
96. Li, T., Ding, C., Jordan, M.: Solving consensus and semi-supervised clustering problems using nonnegative matrix factorization. In: Proceedings of Seventh IEEE International Conference on Data Mining (ICDM 2007), pp. 577–582 (2007)
97. Moon, T.K.: The expectation maximization algorithm. In: IEEE Signal Processing Magazine, pp. 47–60 (1996)
98. Luo, H., Jing, F., Xie, X.: Combining multiple clusterings using information theory based genetic algorithm. IEEE Int. Conf. Comput. Intell. Secur. **1**, 84–89 (2006)
99. Azimi, J., Abdoos, M., Analoui, M.: A new efficient approach in clustering ensembles. Proc. IDEAL'07 Lect. Notes Comput. Sci. **4881**, 395–405 (2007)
100. Chatterjee, S., Mukhopadhyay, A.: Clustering ensemble: a multiobjective genetic algorithm based approach. Procedia Technol. **10**, 443–449 (2013)
101. Ghaemi, R., bin Sulaiman, N., Ibrahim, H., Norwatti, M.: A review: accuracy optimization in clustering ensembles using genetic algorithms. Artif. Intell. Rev. **35**(4), 287–318 (2011)
102. Yan, L., Xin, Y., Tang, W.: Consensus clustering algorithms for asset management in power systems. In: Proceedings of 5th International Conference on Electric Utility Deregulation and Restructuring and Power Technologies (DRPT), pp. 1504–1510 (2015)
103. Manita, G., Khanchel, R., Limam, M.: Consensus functions for cluster ensembles. Appl. Artif. Intell. **26**(6), 598–614 (2012)

104. Kuncheva, L.I., Vetrov, D.P.: Evaluation of stability of k-means cluster ensembles with respect to random initialization. IEEE Trans. Pattern Anal. Mach. Intell. **28**(11), 1798–1808 (2006)

105. Montalvao, J., Canuto, J.: Clustering ensembles and space discretization–a new regard towards diversity and consensus. Pattern Recogn. Lett. **31**(15), 2415–2424 (2010)

106. Zhang, H., Yang, L., Xie, D.: Unsupervised evaluation of cluster ensemble solutions. In: Proceedings of 7th International Conference on Advanced Computational Intelligence, pp. 101–106 (2015)

107. Yeh, C.C., Yang, M.S.: Evaluation measures for cluster ensembles based on a fuzzy generalized Rand index. Appl. Soft Comput. **57**, 225–234 (2017)

108. Alonso-Betanzos, A., Bolón-Canedo, V., Eiras-Franco, C., Morán-Fernández, L., Seijo-Pardo, B.: Preprocessing in high-dimensional datasets. In: Holmes, D., Jain, L. (eds.) Advances in Biomedical Informatics. Intelligent Systems Reference Library, vol. 137, pp. 247–271. Springer, Cham (2018)

109. Irani, K.B.: Multi-interval discretization of continuous-valued attributes for classification learning. In: Proceedings IJCAI, pp. 1022–1029 (1993)

110. Ramírez-Gallego, S., García, S., Mouriño-Talín, H., Martínez-Rego, D., Bolón-Canedo, V., Alonso-Betanzos, A., Benítez, J.M., Herrera, F.: Data discretization: taxonomy and big data challenge. Wiley Interdiscip. Rev. Data Min. Knowl. Discov. **6**, 5–21 (2016)

111. Liu, H., Hussein, F., Tan, C.L., Dash, M.: Discretization: an enabling technique. Data Min. Knowl. Disc. **6**, 393–423 (2002)

112. Sriwanna, K., Boongoen, T., Iam-On, N.: An enhanced univariate discretization based on cluster ensembles. In: Lavangnananda, K., Phon-Amnuaisuk, S., Engchuan, W., Chan, J. (eds.) Intelligent and Evolutionary Systems. Proceedings in Adaptation, Learning and Optimization, pp. 85–98. Springer, Cham (2016)

Chapter 8
Applications of Ensembles Versus Traditional Approaches: Experimental Results

Abstract This chapter presents two different approaches for ensemble feature selection based on the filter model, aiming at achieving a good classification performance together with an important reduction in the input dimensionality. In this manner, we try to overcome the issue of selecting an appropriate method for each problem at hand, as it is usually very dependent on the characteristics of the datasets. The adequacy of using an ensemble of filters instead of a single filter is demonstrated on both synthetic and real data, including the challenging scenario of DNA microarray classification.

Throughout this book, we have discussed all the steps necessary to design an efficient ensemble for feature selection. In this chapter we will see an example with several implementations of ensembles and show how they achieve outstanding results when compared with the standard traditional approach (which is using a single feature selection method to solve a problem).

8.1 The Rationale of the Approach

For years, a typical approach to build an ensemble for feature selection was that the disturbances in the training set due to resampling cause diverse base classifiers to be built or to use different features for each of the base classifiers [1, 2]. Usually, the ensembles found in the literature involving feature selection are based on the idea of applying several feature selection methods in order to distribute the whole set of features into the instances of the classifier [3]. It has to be noted that this method implies that all the features in the training set are exhaustively used.

Part of the content of this chapter was previously published in *Pattern Recognition* (https://doi.org/10.1016/j.patcog.2011.06.006) and *Neurocomputing* (https://doi.org/10.1016/j.neucom.2013.03.067).

© Springer International Publishing AG, part of Springer Nature 2018   139
V. Bolón-Canedo and A. Alonso-Betanzos, *Recent Advances in Ensembles for Feature Selection*, Intelligent Systems Reference Library 147, https://doi.org/10.1007/978-3-319-90080-3_8

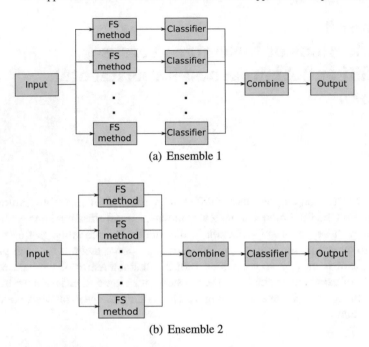

(a) Ensemble 1

(b) Ensemble 2

Fig. 8.1 Implementations of the ensemble

Nevertheless, the purpose of the ensemble presented inhere is different. As commented on Chap. 2, one of the problems of choosing an adequate feature selection method is its variability of results over different datasets. That is, a feature selection method can obtain excellent classification results in a given dataset while performing poorly in another dataset, even in the same domain, depending on the specific properties of the different datasets. Our goal is to achieve a method that reduces the variability of the features selected by the feature selectors in the different classification domains. Therefore, our ensembles are based on the idea of combining several feature selection methods, employing different metrics and performing a feature reduction.

Two distinct general approaches are presented: Ensemble 1 and Ensemble 2 (see Fig. 8.1). The main difference between them is that the former uses several feature selection methods and classifies once for each filter, thus an integration method for the outputs of the classifier is necessary; whilst the later uses several feature selection methods, combines the different subsets returned by each filter, and finally obtains a classification output for this unique subset of features.

8.2 The Process of Selecting the Methods for the Ensemble

As stated in Chap. 2, feature selection algorithms designed with different evaluation criteria broadly fall into three categories: the filter model, the embedded model and the wrapper model. The objective of the ensemble presented here is that it can be applied to high-dimensional data, such as DNA microarray, so the wrapper model is discarded because it could not generalize adequately. Therefore, in a first stage, filters and embedded methods were chosen to perform a previous study, paving the way for its application to the ensemble.

As the goal is to choose methods based on different metrics, five filters and two embedded methods were tested over five synthetic data sets under different situations: increasing number of irrelevant features and the insertion of noise in the inputs, as well as the inclusion of correlated features. Both filter and embedded methods were described in detail in Chap. 2 and it has to be noted that three of them (CFS, consistency-based and INTERACT) provide a subset of features, whereas the remaining four (Information Gain, ReliefF, SVM-RFE and FS-P) provide features ordered according to their relevance (a ranking of features).

In order to determine the effectiveness of each one of the feature selection methods mentioned above at different situations, several widely-used synthetic datasets were employed (see Chap. 2): the LED dataset, the CorrAL dataset and the XOR-100 dataset.

Table 8.1 shows the score for each feature selection method over each scenario and also an overall score for each method (last column). This score is defined as:

$$score = \left[\frac{R_s}{R_t} - \frac{I_s}{I_t} \right] \times 100,$$

where R_s is the number of relevant features selected, R_t is the total number of relevant features, I_s is the number of irrelevant features selected and I_t is the total number of irrelevant features. Notice that 100 is the desired value for this score and negative values indicate a high selection of irrelevant features.

Table 8.1 Score for each feature selection method tested

Method	CorrAL	CorrAL-100	XOR-100	Led-25	Led-100	Average
CFS	50.00	94.00	46.00	71.50	71.33	66.57
Consistency	50.00	94.00	46.00	68.00	64.00	64.40
INTERACT	25.00	92.00	47.00	66.67	73.50	60.83
InfoGain	0.00	88.00	−1.00	66.33	70.00	44.67
ReliefF	50.00	88.00	95.00	78.17	82.50	78.73
SVM-RFE	50.00	59.00	−15.00	22.83	25.33	27.93
FS-P	0.00	43.00	−9.00	72.00	70.67	35.33

As can be seen in Table 8.1, the two embedded methods (SVM-RFE and FS-P) achieve the poorest scores. As SVM-RFE achieved the worst result, we decided not to use it in our ensemble. Focusing on the filters, although ReliefF obtained the best average, CFS, Consistency and INTERACT also showed a good performance. Information Gain obtained the poorest results of the filters methods, and similar to those obtained by FS-P. However, since Information Gain performs better than FS-P and bearing in mind the higher computational cost of the embedded methods, FS-P is discarded. Thus, all the five filters were selected to conform our ensemble.

8.3   Two Filter Ensemble Approaches

As mentioned before, when dealing with ensemble feature selection, a typical practice is to use different features for each of the base classifiers. However, with our ensemble, not all the features have to be necessarily employed, since the idea is to apply several filters based on different metrics so as to have a diverse set of selections. By using this ensemble of filters, the user is released from the task of choosing an adequate filter for each scenario, because this approach obtains acceptable results independently of the characteristics of the data. Among the broad suite of filters available in the literature, five filters were selected according to a study performed in Sect. 8.2, all of them based on different metrics. Two distinct general approaches are proposed: Ensemble 1 and Ensemble 2 (see Fig. 8.1). The main difference between them are that the former uses several filters and classifies once for each filter, as an integration method for the outputs of the classifier is necessary, whilst the later uses several filters, combines the different subsets returned by each filter, and finally obtains a classification output for this unique subset of features.

8.3.1   Ensemble 1

This is a more classic approach, consisting of an ensemble of classifiers including a previous stage of feature selection (see Fig. 8.1a). Particularly, each one of the F filters selects a subset of features and this subset is used for training a given classifier. Therefore, there will be as many outputs as filters were employed in the ensemble (F). Due to the different metric the filters are based on, they select different sets of features leading to classifier outputs that could be contradictory, so a combination method becomes necessary (see Chap. 5). Note that in each execution F filters and only one classifier are used, but the classifier is trained F times (once for each filter). The pseudo-code is shown in Algorithm 8.1. Different variants of this philosophy will be implemented regarding the combination of the F outputs. Two different methods are considered, producing two implementations of Ensemble 1. The first uses majority vote or (E1-mv), where for a particular instance, each classifier votes for a class and the class with the greatest number of votes is considered the output class. The second implementation (E1-max) stores the probability with which an instance has been assigned to a class and then use the max rule to decide the output (see Chap. 5).

Algorithm 8.1: Pseudo-code for *Ensemble 1*

Data: $F \leftarrow$ number of filters

Result: P $\leftarrow$ classification prediction
1 **for** *each f from 1 to F* **do**
2   Select attributes A using filter f
3   Build classifier C_f with the selected attributes A
4   Obtain prediction P_f from classifier
   end
5 Apply a combination method over predictions $P_1 \ldots P_f$
6 Obtain prediction P

Another variation in the basic scheme of Ensemble 1 comes from thinking that instead of using the same classifier for all five filters, there might be classifiers more suitable for certain feature selection methods. In fact, in Chap. 2 it was stated that CFS, Consistency-based, INTERACT and InfoGain select a small number of relevant features, whilst ReliefF is very effective at removing redundancy. On the other hand, k-NN and SVM deteriorate their performance when irrelevant features are present whereas naive Bayes is robust with respect to irrelevant features but deteriorates with redundant ones. In this situation, we propose to try an ensemble which uses naive Bayes together with ReliefF and k-NN with the remaining filters (E1-nk) and another which uses again naive Bayes together with ReliefF and SVM with the remaining filters (E1-ns). Both these configurations can be seen in Fig. 8.2.

8.3.2 Ensemble 2

This approach consists of combining the subsets selected by each one of the F filters obtaining only one subset of features. This method has the advantage of not requiring a combiner method in order to obtain the class prediction. On the contrary, it needs a method to combine the features returned by each F filter, as can be seen in Fig. 8.1b, since it only employs one classifier. Strategies such as the union or the intersection of the subsets usually lead to poor results due to the redundancy induced by the union or the extremely aggressive reduction in the set of features produced by the intersection. Thus, as can be seen in Algorithm 8.2, we combine the subsets of features so as to add to the final subset only those subsets capable of outperforming the classification accuracy in the training set (see Sect. 5.2 in Chap. 5).

The complexity of these two ensembles depends on the machine learning algorithms used. Let K and J be the complexities of the feature selection and the data

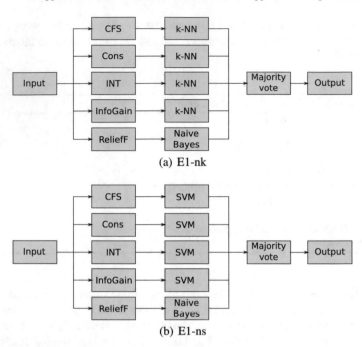

(a) E1-nk

(b) E1-ns

Fig. 8.2 Configurations of Ensemble 1: E1-nk and E1-ns

Algorithm 8.2: Pseudo-code for *Ensemble2*

Data: $F \leftarrow$ number of filters

Result: P $\leftarrow$ classification prediction
1 **for** *each f from 1 to F* **do**
2 | Select attributes A_f using filter f
   end
3 $A = A_f$
4 $baseline =$ classifying subset A with classifier C
5 **for** *each f from 2 to F* **do**
6 | $A_{aux} = A \cup A_f$
7 | $accuracy =$ classifying subset A_{aux} with classifier C
8 | **if** *accuracy > baseline* **then**
9 | | $A = A_{aux}$
10 | | $baseline = accuracy$
   | **end**
  end
11 Build classifier C with the selected attributes A
12 Obtain prediction P

---

mining algorithms, respectively, and F the number of filters used in the ensembles. The complexity will be $F \max(K, J)$. Since the idea of both Ensemble 1 and Ensemble 2 is to use a small number of filters, compared with the number of features or samples of the datasets, F can be considered negligible and it can be said that the complexity of these ensembles is determined by the method with the higher complexity (either K or J). Therefore, it is not more computationally complex than the filters employed alone.

8.4 Experimental Setup

Although the final goal of a feature selection method is to test its effectiveness over a real dataset, a first step showing performance over synthetic data follows. The main advantage of artificial scenarios is the knowledge of the set of optimal features that must be selected, thus the degree of closeness to any of these solutions can be assessed in a confident way. The LED problem (see Chap. 2) has been chosen as the synthetic dataset to test the ensembles. It is a simple classification task that consists of, given the active leds on a seven segments display, identifying the digit that the display is representing. Therefore, the classification task to be solved is described by 7 binary attributes and 10 possible classes available. In particular, it will be used the dataset Led100, which consists of 50 samples and 100 features, where 92 irrelevant attributes (with random binary values) have been added to the 7 relevant ones.

Then, to check if the behavior displayed by the ensembles over the synthetic dataset can be extrapolated to the real world, 5 real classical datasets were chosen, which can be consulted in Table 8.2. This suite of datasets represents different problematic that can appear in real data, such as non-linearity (Madelon) or high imbalance of the classes (Ozone). These datasets have only available a training dataset, so a 10-fold cross-validation will be performed. Finally, and in order to widen the scope of this research, the proposed ensemble will be also tested over a challenging scenario: DNA microarray data. These type of datasets poses an enormous challenge for feature selection researchers due to their high number of gene expression and the small sample size. Seven well-known binary microarray datasets are considered: Colon, DLBCL, CNS, Leukemia, Prostate, Lung and Ovarian. Those datasets originally divided into training and test sets were maintained, whereas, for the sake of comparison, datasets with only training set were randomly divided using the common rule 2/3 for training and 1/3 for testing. This division introduces a more challenging scenario, since in some datasets, the distribution of the classes in the training set differs from the one in the test set. Table 8.3 depicts the number of attributes and samples and also the distribution of the binary classes, i.e. the percentage of binary labels in the datasets, showing if the data is unbalanced.

While three of the filters which form part of the ensemble return a feature subset (CFS, Consistency-base and INTERACT), the other two (ReliefF and Information Gain) are ranker methods, so it is necessary to establish a threshold in order to obtain a subset of features. Initial experiments on microarray data showed that for most of

Table 8.2 Dataset description for binary classic datasets

Dataset	Features	Samples	Distribution %
Madelon	500	2400	50 – 50
Mushrooms	112	8124	48 – 52
Ozone	72	2536	97 – 3
Spambase	57	4601	61 – 39
Splice	60	1000	48 – 52

Table 8.3 Dataset description for binary microarray datasets

Dataset	Features	Samples	Distribution %
CNS	7129	60	35 – 65
Colon	2000	62	35 – 65
DLBCL	4026	47	49 – 51
Leukemia	7129	72	34 – 66
Lung	12533	181	17 – 83
Ovarian	15154	253	36 – 64
Prostate	12600	136	43 – 57

the datasets, the subset filters selected a number of features between 25 and 50. For the sake of fairness, the rankers were forced to select a number of features similar to the cardinality obtained by the other type of filters. Several experiments were carried out with 25 and 50 features. As performance did not improve using 50 features with respect to 25, we have decided to force these ranker methods to obtain subsets with 25 features. Finally, to test the performance of the different ensembles of filters proposed it is necessary to use a classifier which provides classification accuracy as a measure of adequacy of the method. For this purpose, four well-known classifiers were chosen: C4.5, naive Bayes, k-NN and SVM (see descriptions in Chap. 6).

8.5   Experimental Results

In this section the results obtained after applying our ensembles will be shown. To sum up, five ensemble approaches will be tested: *E1-mv*, which is Ensemble 1 using majority vote as combination method; *E1-max*, which is Ensemble 1 using the max rule as combination method; *E1-nk*, which is Ensemble 1 with specific classifiers naive Bayes and k-NN; *E1-ns*, which is Ensemble 1 with specific classifiers naive Bayes and SVM and *E2*, which is Ensemble 2.

Table 8.4 Error results over the synthetic dataset Led100 as well as number of relevant and irrelevant features selected

		No. rel.	No. irrel.	C4.5	NB	k-NN	SVM
Ensembles	E1-mv	6	2	6.25	0.00	0.00	6.25
	E1-max	6	2	6.25	0.00	0.00	6.25
	E1-nk	6	2	0.00	0.00	0.00	0.00
	E1-ns	6	2	6.25	6.25	6.25	6.25
	E2	6	0	6.25	0.00	0.00	6.25
Filters	CFS	6	0	6.25	0.00	0.00	6.25
	Cons	5	0	6.25	0.00	0.00	6.25
	INT	6	0	6.25	0.00	0.00	6.25
	IG	6	1	6.25	12.50	12.50	0.00
	ReliefF	5	2	18.75	31.25	31.25	18.75

8.5.1 Results on Synthetic Data

Table 8.4 shows the results obtained by the ensembles and the filters alone over the synthetic dataset Led100. The number of relevant features selected (notice that the optimal is 7), the number of irrelevant features selected (notice that the maximum is 92) and the test classification error are exhibited, after randomly dividing the dataset using the common rule 2/3 for training and 1/3 for testing. Please note that in this concrete case, the number of features the rankers were forced to select was 7, corresponding with the optimal number of relevant features.

The results demonstrate the adequacy of the proposed ensembles, since they matched or improved upon the results achieved by the filters alone. Focusing on the features selected, it is important to note that although the theoretical number of relevant features is 7 (one for each led segment), there are two segments that are not relevant for distinguishing among the 10 numbers. For this reason, the consistency filter was able to correctly classify all the instances using only 5 out of the 7 theoretical relevant features. The reader should also notice that the features selected by the four Ensemble 1 approaches are the union of the features selected by each one of the filters. Therefore it is more informative to focus on the classification error. According to this measure, it is easy to see that the ensembles take advantage of the filters which work correctly on a dataset and discard the influence of those which do not (IG and ReliefF, in the dataset at hand).

8.5.2 Results on Classical Datasets

After verifying on synthetic data that our proposed ensembles behave in a confident way, the next step is to evaluate their performance on real classical datasets. Five datasets were chosen for this task (Ozone, Spambase, Mushrooms, Splice and Madelon), which can be consulted in Table 8.2. In this case, a 10-fold cross validation will be used and the results obtained are depicted in Table 8.5: average test classification error along with the average number of features required to train the model. In the case of the classifier alone, it uses the whole set of features. When using this type of validation with several repetitions, the use of statistical inference for analyzing the results is a crucial and necessary task in an investigation.

For this reason, a Kruskal–Wallis test was applied to check if there are significant differences among the medians for each method for a level of significance $\alpha = 0.05$. If differences among the medians were found, a multiple comparison procedure (Tukey's) was applied to find the simplest approach whose classification error is not significantly different for the approach with the lowest error (labeled with a cross in the tables).

For all datasets and classifiers, one of the five ensembles presented here obtains the lowest error, showing the adequacy of the ensemble approach in these standard datasets. It is necessary to note that Ozone is an extremely unbalanced dataset (see Table 8.2) and by assigning all the samples to the majority class, an error of 2.88% could be obtained. None of the methods tested was able to improve this result, although some methods matched it. An oversampling technique was applied over this difficult dataset but due to its extremely high imbalance, no improvement was obtained. On the other hand, it is worth mentioning that E1-nk achieves a promising result over Madelon dataset. In fact, it reduces the test error up to 33% compared with the classifier alone and up to 24% compared with CFS filter, using only 8% of the total features (see results for SVM).

These results are not easy to analyze since the classifier plays a crucial role and provides a very different classification error even with the same set of features. There are several cases found in Table 8.5 that confirm this fact, for example: ReliefF over Ozone dataset achieves an error of 2.88% according to SVM whilst naive Bayes classifier raises the error up to 29.06%; and the consistency filter over Madelon dataset increases its error from 9.08–33.42% using k-NN and SVM, respectively.

Table 8.6 displays the average of test error for each dataset and method, independently of the classifier, which should help to clarify which one is the best method for a given dataset. E1-nk is the method which is significantly better in the maximum number of datasets (3), followed by E2 and the consistency filter.

Table 8.7 depicts the average of test error for each method and classifier, independently of the dataset. In this case it makes no sense to perform a statistical study since the results achieved by the classifiers over different datasets are very different. We can see that for these kinds of datasets, the best option is to use E1-max combined with C4.5 classifier. It is also worth noting that for the remainder of classifiers tested, it is always one of the ensembles which achieves the best results, outperforming the results obtained by the filters alone. In the next section we will see if the methods tested exhibit the same behavior when dealing with an extremely complex scenario: DNA microarray data classification.

Table 8.5 Test classification error after 10 fold cross-validation for classical datasets, the number in parenthesis is the number of features selected by the method. Those methods whose average test classification results are not significantly worse than the best are labeled with a cross (†)

		Method	Ozone	Spambase	Mushrooms	Splice	Madelon
C4.5	Ensembles	E1-mv	$3.35 (43)^\dagger$	$5.76 (44)^\dagger$	$0.00 (42)^\dagger$	$4.10 (32)^\dagger$	16.17 (38)
		E1-max	$3.23 (43)^\dagger$	$5.69 (44)^\dagger$	$0.00 (42)^\dagger$	$4.10 (32)^\dagger$	15.42 (38)
		E1-nk	4.18 (43)	8.22 (44)	$0.00 (42)^\dagger$	15.10 (32)	$9.08 (38)^\dagger$
		E1-ns	$2.88 (43)^\dagger$	11.50 (44)	1.13 (42)	20.70 (32)	33.25 (38)
		E2	$3.71 (21)^\dagger$	$6.74 (31)^\dagger$	$0.00 (22)^\dagger$	$5.80 (18)^\dagger$	$14.21 (25)^\dagger$
		C4.5	4.61	$6.67^\dagger$	$0.00^\dagger$	$6.00^\dagger$	19.88
	Filters	CFS	$3.71 (17)^\dagger$	7.43 (15)	1.48 (8)	$6.00 (12)^\dagger$	19.33 (8)
		Cons	$3.08 (4)^\dagger$	$7.35 (25)^\dagger$	0.15 (10)	$5.50 (12)^\dagger$	17.04 (13)
		INT	$3.82 (19)^\dagger$	$7.32 (29)^\dagger$	$0.00 (10)^\dagger$	$4.90 (15)^\dagger$	17.04 (13)
		IG	$3.67 (25)^\dagger$	7.43 (25)	$0.00 (25)^\dagger$	$5.90 (25)^\dagger$	18.33 (25)
		ReliefF	$3.86 (25)^\dagger$	8.80 (25)	$0.00 (25)^\dagger$	$5.90 (25)^\dagger$	$14.08 (25)^\dagger$
Naive bayes	Ensembles	E1-mv	21.06 (43)	14.39 (44)	6.84 (42)	$17.60 (32)^\dagger$	30.17 (38)
		E1-max	21.14 (43)	13.39 (44)	6.87 (42)	$16.70 (32)^\dagger$	29.92 (38)
		E1-nk	$4.18 (43)^\dagger$	$8.22 (44)^\dagger$	$0.00 (42)^\dagger$	$15.10 (32)^\dagger$	$9.08 (38)^\dagger$
		E1-ns	$2.88 (43)^\dagger$	$11.50 (44)^\dagger$	$1.13 (42)^\dagger$	$20.70 (32)^\dagger$	33.25 (38)
		E2	20.98 (17)	15.24 (32)	$1.23 (14)^\dagger$	$16.50 (27)^\dagger$	30.08 (15)
		NB	28.98	20.47	6.81	$16.40^\dagger$	31.38
	Filters	CFS	20.98 (17)	21.23 (15)	$1.48 (8)^\dagger$	$17.90 (12)^\dagger$	30.29 (8)
		Cons	$6.94 (4)^\dagger$	$11.82 (25)^\dagger$	4.89 (10)	$17.50 (12)^\dagger$	29.92 (13)
		INT	20.58 (19)	16.82 (29)	5.45 (10)	$16.00 (15)^\dagger$	29.92 (13)
		IG	26.10 (25)	$11.63 (25)^\dagger$	6.95 (25)	$17.10 (25)^\dagger$	30.42 (25)
		ReliefF	29.06 (25)	29.41 (25)	6.60 (25)	$16.30 (25)^\dagger$	30.08 (25)
k-NN	Ensembles	E1-mv	$3.51 (43)^\dagger$	$7.93 (44)^\dagger$	$0.00 (42)^\dagger$	$15.50 (32)^\dagger$	$8.92 (38)^\dagger$
		E1-max	$3.51 (43)^\dagger$	$7.93 (44)^\dagger$	$0.00 (42)^\dagger$	$15.50 (32)^\dagger$	$8.92 (38)^\dagger$
		E1-nk	4.18 (43)	$8.22 (44)^\dagger$	$0.00 (42)^\dagger$	$15.10 (32)^\dagger$	$9.08 (38)^\dagger$
		E1-ns	$2.88 (43)^\dagger$	11.50 (44)	1.13 (42)	20.70 (32)	33.25 (38)
		E2	4.85 (28)	10.76 (19)	$0.00 (19)^\dagger$	20.20 (15)	$9.08 (13)^\dagger$

(continued)

Table 8.5 (continued)

		Method	Ozone	Spambase	Mushrooms	Splice	Madelon
		k-NN	4.73	$9.09^{\dagger}$	$0.00^{\dagger}$	30.80	41.04
	Filters	CFS	4.97 (17)	11.50 (15)	1.75 (8)	20.50 (12)	13.38 (8)
		Cons	$3.23 (4)^{\dagger}$	10.43 (25)	$0.00 (10)^{\dagger}$	19.40 $(12)^{\dagger}$	$9.08 (13)^{\dagger}$
		INT	4.85 (19)	10.61 (29)	$0.00 (10)^{\dagger}$	19.10 $(15)^{\dagger}$	$9.08 (13)^{\dagger}$
		IG	4.61 (25)	10.32 (25)	$0.00 (25)^{\dagger}$	24.70 (25)	23.08 (25)
		ReliefF	4.30 (25)	12.54 (25)	$0.00 (25)^{\dagger}$	21.60 (25)	10.96 $(25)^{\dagger}$
SVM	Ensembles	E1-mv	$2.88 (43)^{\dagger}$	12.00 (44)	1.08 (42)	20.80 (32)	33.46 (38)
		E1-max	$2.88 (43)^{\dagger}$	12.00 (44)	1.08 (42)	20.80 (32)	33.46 (38)
		E1-nk	4.18 (43)	$8.22 (44)^{\dagger}$	$0.00 (42)^{\dagger}$	15.10 $(32)^{\dagger}$	$9.08 (38)^{\dagger}$
		E1-ns	$2.88 (43)^{\dagger}$	11.50 (44)	1.13 (42)	20.70 $(32)^{\dagger}$	33.25 (38)
		E2	$2.88 (17)^{\dagger}$	10.28 $(41)^{\dagger}$	$0.00 (19)^{\dagger}$	19.70 $(25)^{\dagger}$	33.58 (16)
		SVM	$2.88^{\dagger}$	$9.59^{\dagger}$	$0.00^{\dagger}$	$20.40^{\dagger}$	42.25
	Filters	CFS	$2.88 (17)^{\dagger}$	13.17 (15)	1.66 (8)	21.00 (12)	34.04 (8)
		Cons	$2.88 (4)^{\dagger}$	12.50 (25)	$0.00 (10)^{\dagger}$	21.90 (12)	33.42 (13)
		INT	$2.88 (19)^{\dagger}$	11.76 (29)	2.26 (10)	21.60 (15)	33.42 (13)
		IG	$2.88 (25)^{\dagger}$	12.08 (25)	1.86 (25)	20.00 $(25)^{\dagger}$	33.50 (25)
		ReliefF	$2.88 (25)^{\dagger}$	13.50 (25)	$0.10 (25)^{\dagger}$	19.30 $(25)^{\dagger}$	33.67 (25)

Table 8.6 Average of test error for classical datasets focusing on the dataset. Those methods whose average test classification results are not significantly worse than the best are labeled with a cross (†)

		Ozone	Spambase	Mushrooms	Splice	Madelon
Ensembles	E1-mv	7.70	10.02	1.98	$14.50^{\dagger}$	22.18
	E1-max	7.69	$9.75^{\dagger}$	1.99	$14.28^{\dagger}$	21.93
	E1-nk	4.18	$8.22^{\dagger}$	$0.00^{\dagger}$	15.10	$9.08^{\dagger}$
	E1-ns	$2.88^{\dagger}$	11.50	1.13	20.70	33.25
	E2	8.10	10.75	$0.31^{\dagger}$	$15.55^{\dagger}$	21.74
	Classif	10.30	11.45	1.70	$18.40^{\dagger}$	33.64
Filters	CFS	8.13	13.33	1.59	$16.35^{\dagger}$	24.26
	Cons	$4.03^{\dagger}$	10.52	1.26	$16.07^{\dagger}$	22.36
	INT	8.03	11.63	1.93	$15.40^{\dagger}$	22.36
	IG	9.32	10.37	2.20	$16.93^{\dagger}$	26.33
	ReliefF	10.03	16.06	1.67	$15.78^{\dagger}$	22.20

Table 8.7 Average of test error for classical datasets focusing on the classifier

		C4.5	NB	k-NN	SVM
Ensembles	E1-mv	5.88	18.01	7.17	14.04
	E1-max	5.69	17.60	7.17	14.04
	E1-nk	7.32	7.32	7.32	7.32
	E1-ns	13.89	13.89	13.89	13.89
	E2	6.09	16.81	8.98	13.29
	Classif	7.43	20.81	17.13	15.02
Filters	CFS	7.59	18.38	10.42	14.55
	Cons	6.62	14.21	8.43	14.14
	INT	6.62	17.76	8.73	14.38
	IG	7.07	18.44	12.54	14.06
	ReliefF	6.53	22.29	9.88	13.89

8.5.3 Results on Microarray Data

The last step for testing our ensembles is to evaluate them on a difficult scenario such as DNA microarray classification, where the number of features is much higher than the number of samples. Remind that, in this case, a division in training and test sets is assumed (see Sect. 8.4).

Table 8.8 exhibits the results over all the seven microarray datasets considered for the classifiers used in the experiments. Along with the error test achieved, one can see the number of features required to train the model. In the case of the classifier alone, it uses the whole set of features. For all the four classifiers employed, one of the five ensemble approaches proposed achieves the lowest error, except for Colon dataset with C4.5 and k-NN. Although the number of features is higher using ensembles than filters alone, it is insignificant when compared with the difference in feature number regarding the complete original feature set.

In order to summarize, Table 8.9 shows the results on average. The best result on average for all datasets and classifiers is obtained by E1-mv and E1-max combined with SVM classifier, which happens to be a frequently used and appropriate classifier for DNA microarray classification [4–6]. As in the classical datasets case, again one of the ensembles achieves always the best result for each classifier. It also should to be noted that there is a slight difference between using the max rule (E1-max) or majority vote (E1-mv) as union method in Ensemble 1. E1-max only appears to be better for naive Bayes classifier, but as it does not produce deterioration for any classifier, it is considered to be a better choice than E1-mv. For all these reasons, we recommend to use E1-max combined with SVM classifier when dealing with DNA microarray data.

Table 8.8 Test classification error for microarray datasets, the number in parenthesis is the number of features selected by the method. Best error for each dataset is highlighted in bold face

	Method	Colon	DLBCL	CNS	Leukemia	Prostate	Lung	Ovarian
C4.5 — Ensembles	E1-mv	15.00 (58)	13.33 (73)	50.00 (95)	8.82 (63)	73.53 (126)	18.12 (55)	**0.00** (67)
	E1-max	15.00 (58)	13.33 (73)	50.00 (95)	8.82 (63)	73.53 (126)	18.12 (55)	**0.00** (67)
	E1-nk	20.00 (58)	13.33 (73)	40.00 (95)	**5.88** (63)	67.65 (126)	**0.00** (55)	**0.00** (67)
	E1-ns	20.00 (58)	**6.67** (73)	**30.00** (95)	11.76 (63)	**29.41** (126)	1.34 (55)	**0.00** (67)
	E2	15.00 (34)	13.33 (47)	50.00 (60)	8.82 (36)	73.53 (89)	18.12 (40)	**0.00** (37)
	C4.5	**10.00**	13.33	40.00	8.82	73.53	18.12	1.19
C4.5 — Filters	CFS	15.00 (19)	13.33 (47)	50.00 (60)	8.82 (36)	73.53 (89)	18.12 (40)	**0.00** (37)
	Cons	15.00 (3)	13.33 (2)	50.00 (3)	8.82 (1)	76.47 (4)	18.12 (1)	**0.00** (3)
	INT	15.00 (16)	13.33 (36)	45.00 (47)	8.82 (36)	73.53 (73)	18.12 (40)	1.19 (27)
	IG	30.00 (25)	13.33 (25)	50.00 (25)	8.82 (25)	70.59 (25)	10.07 (25)	**0.00** (25)
	ReliefF	15.00 (25)	13.33 (25)	35.00 (25)	8.82 (25)	67.65 (25)	2.68 (25)	1.19 (25)
Naive bayes — Ensembles	E1-mv	20.00 (58)	**6.67** (73)	45.00 (95)	8.82 (63)	73.53 (126)	**0.00** (55)	1.19 (67)
	E1-max	20.00 (58)	**6.67** (73)	40.00 (95)	8.82 (63)	73.53 (126)	**0.00** (55)	1.19 (67)
	E1-nk	20.00 (58)	13.33 (73)	40.00 (95)	**5.88** (63)	67.65 (126)	**0.00** (55)	**0.00** (67)
	E1-ns	20.00 (58)	**6.67** (73)	**30.00** (95)	11.76 (63)	**29.41** (126)	1.34 (55)	**0.00** (67)
	E2	25.00 (34)	**6.67** (47)	35.00 (60)	**5.88** (36)	73.53 (89)	**0.00** (40)	2.38 (37)
	NB	30.00	**6.67**	40.00	11.76	73.53	4.70	11.90
Naive bayes — Filters	CFS	**10.00** (19)	**6.67** (47)	**30.00** (60)	**5.88** (36)	73.53 (89)	**0.00** (40)	2.38 (37)
	Cons	15.00 (3)	13.33 (2)	45.00 (3)	8.82 (1)	67.65 (4)	14.09 (1)	**0.00** (3)
	INT	15.00 (16)	**6.67** (36)	35.00 (47)	**5.88** (36)	73.53 (73)	**0.00** (40)	**0.00** (27)
	IG	15.00 (25)	**6.67** (25)	45.00 (25)	8.82 (25)	73.53 (25)	0.67 (25)	2.38 (25)
	ReliefF	20.00 (25)	**6.67** (25)	40.00 (25)	8.82 (25)	79.41 (25)	**0.00** (25)	2.38 (25)

(continued)

Table 8.8 (continued)

	Method	Colon	DLBCL	CNS	Leukemia	Prostate	Lung	Ovarian
k-NN								
Ensembles	E1-mv	20.00 (58)	13.33 (73)	35.00 (95)	14.71 (63)	67.65 (126)	**0.00** (55)	**0.00** (67)
	E1-max	20.00 (58)	13.33 (73)	35.00 (95)	14.71 (63)	67.65 (126)	**0.00** (55)	**0.00** (67)
	E1-nk	20.00 (58)	13.33 (73)	40.00 (95)	**5.88** (63)	67.65 (126)	**0.00** (55)	**0.00** (67)
	E1-ns	20.00 (58)	**6.67** (73)	**30.00** (95)	11.76 (63)	**29.41** (126)	1.34 (55)	**0.00** (67)
	E2	50.00 (34)	13.33 (47)	35.00 (60)	14.71 (36)	67.65 (89)	**0.00** (40)	**0.00** (37)
	k-NN	**5.00**	26.67	45.00	29.41	47.06	2.01	7.14
Filters	CFS	20.00 (19)	13.33 (47)	35.00 (60)	14.71 (36)	67.65 (89)	**0.00** (40)	**0.00** (37)
	Cons	15.00 (3)	26.67 (2)	35.00 (3)	8.82 (1)	73.53 (4)	18.12 (1)	**0.00** (3)
	INT	20.00 (16)	13.33 (36)	40.00 (47)	14.71 (36)	67.65 (73)	**0.00** (40)	**0.00** (27)
	IG	15.00 (25)	**6.67** (25)	30.00 (25)	**5.88** (25)	58.82 (25)	1.34 (25)	3.57 (25)
	ReliefF	15.00 (25)	**6.67** (25)	40.00 (25)	17.65 (25)	70.59 (25)	1.34 (25)	**0.00** (25)
SVM								
Ensembles	E1-mv	**15.00** (58)	**6.67** (73)	**30.00** (95)	14.71 (63)	**2.94** (126)	1.34 (55)	**0.00** (67)
	E1-max	**15.00** (58)	**6.67** (73)	**30.00** (95)	14.71 (63)	**2.94** (126)	1.34 (55)	**0.00** (67)
	E1-nk	20.00 (58)	13.33 (73)	40.00 (95)	**5.88** (63)	67.65 (126)	**0.00** (55)	**0.00** (57)
	E1-ns	20.00 (58)	**6.67** (73)	**30.00** (95)	11.76 (63)	29.41 (126)	1.34 (55)	**0.00** (57)
	E2	20.00 (34)	**6.67** (47)	35.00 (60)	11.76 (36)	**2.94** (89)	1.34 (40)	**0.00** (37)
	SVM	25.00	13.33	**30.00**	14.71	47.06	0.67	**0.00**
Filters	CFS	**15.00** (19)	**6.67** (47)	35.00 (60)	11.76 (36)	**2.94** (89)	1.34 (40)	**0.00** (37)
	Cons	20.00 (3)	13.33 (2)	35.00 (3)	20.59 (1)	73.53 (4)	18.12 (1)	**0.00** (3)
	INT	**15.00** (16)	13.33 (36)	40.00 (47)	11.76 (36)	29.41 (73)	1.34 (40)	**0.00** (27)
	IG	**15.00** (25)	**6.67** (25)	35.00 (25)	11.76 (25)	**2.94** (25)	0.67 (25)	1.19 (25)
	ReliefF	**15.00** (25)	**6.67** (25)	**30.00** (25)	17.65 (25)	5.88 (25)	2.01 (25)	**0.00** (25)

Table 8.9 Average of test error

		C4.5	NB	k-NN	SVM
Ensembles	E1-mv	25.54	22.17	21.53	**10.09**
	E1-max	25.54	21.46	21.53	**10.09**
	E1-nk	20.98	20.98	20.98	20.98
	E1-ns	**14.17**	**14.17**	**14.17**	14.17
	E2	25.54	20.49	25.81	11.10
	Classifier	23.57	25.51	23.18	18.68
Filters	CFS	25.54	18.35	21.53	10.39
	Cons	25.96	23.41	25.31	25.80
	INT	25.00	19.44	22.24	15.83
	IG	26.12	21.72	17.33	10.46
	ReliefF	20.52	22.47	21.61	11.03

Table 8.10 Average of test error after applying SMOTE for datasets Colon, CNS, Leukemia and Ovarian

		C4.5	NB	k-NN	SVM
Ensembles	E1-mv	**13.46**	17.50	**13.68**	**15.96**
	E1-max	**13.46**	17.50	**13.68**	**15.96**
	E1-nk	16.47	16.47	16.47	16.47
	E1-ns	18.46	18.46	18.46	18.46
	E2	16.25	**15.82**	24.93	20.44
	Classifier	17.09	26.36	25.98	17.43
Filters	CFS	16.25	**15.82**	24.93	20.44
	Cons	17.50	17.21	15.96	18.46
	INT	14.71	16.47	21.18	20.44
	IG	15.00	17.80	14.86	18.54
	ReliefF	16.25	18.32	17.21	18.16

8.5.4   The Imbalance Problem

Four of the microarray datasets considered in the experiments in this chapter presented the so-called imbalance problem (Colon, CNS, Leukemia and Ovarian; see Table 8.3). A dataset is considered unbalanced when the classification categories are not approximately equally represented [7].

To overcome this issue, the SMOTE method [8] is applied after the feature selection process in the datasets that show imbalance in the training set. For the sake of brevity, only the average of test error will be shown in Table 8.10. E1-mv and E1-max obtained again the lowest error combined with C4.5 classifier. As the results obtained are better using SMOTE, the adequacy of this oversampling technique when combined with ensemble techniques is confirmed.

8.6 Summary

In this chapter we have presented two general approaches for ensembles for feature selection applied to different real-life datasets. Ensemble 1 classifies as many times as there are filters, whereas Ensemble 2 classifies only once with the result of joining the different subsets selected by the filters. For Ensemble 1, two methods for combining the outputs of the classifiers were studied (majority vote and max rule), as well as the possibility of using an adequate specific classifier for each filter. A total of five different implementations of the two approaches of ensemble were presented, tested in the first place over synthetic data. Results showed the adequacy of the proposed methods on this controlled scenario since they selected the correct features. The next step was to apply these approaches over 5 UCI classical datasets. Experimental results demonstrated that one of the ensembles (E1-max) combined with C4.5 classifier was the best option when dealing with this type of dataset. Finally, the ensemble configurations were tested over 7 DNA microarray data. These are extremely challenging datasets because of their high number of input features and small sample size, where feature selection becomes indispensable. It turned out that using an ensemble was again the best option. Specifically, the best performance was achieved again with E1-max but this time combined with SVM classifier. It should be noted that some of these datasets presented a high imbalance of the data. To overcome this problem, an oversampling method was applied after the feature selection process. The result was that once again one of the ensembles achieved the best performance, and that this was even better than the one obtained with no preprocessing, showing the adequacy of the ensemble combined with over-sampling methods. Thus, the appropriateness of using an ensemble instead of a single filter remained demonstrated, considering that for all scenarios tested, the ensemble was always the more successful solution.

Regarding the different implementations of the ensemble tested, several conclusions can be drawn. There is a slight difference between the two combiner methods employed with Ensemble 1 (majority vote and max rule), although the second one obtained the best performance. Among the different classifiers chosen for this study, it appeared that the type of data to be classified determines significantly the error achieved, so it is responsibility of the user to know which classifier is more suitable for a given type of data. We recommend using E1-max with C4.5 when classifying classical datasets (with more samples than features) and E1-max with SVM when dealing with microarray dataset (with more features than samples). In complete ignorance of the particulars of the data, we suggest using E1-ns, which releases the user from the task of choosing a specific classifier.

References

1. Ho, T.K.: The random subspace method for constructing decision forests. IEEE Trans. Pattern Anal. Mach. Intell. **20**(8), 832–844 (1998)
2. Opitz, D.W.: Feature selection for ensembles. In: Proceedings of the National Conference on Artificial Intelligence, pp. 379–384 (1999)
3. Pradhananga, N.: Effective linear-time feature selection, Ph.D. thesis (2007)
4. Mukherjee, S., Tamayo, P., Slonim, D., Verri, A., Golub, T., Mesirov, J., Poggio, T.: Support vector machine classification of microarray data, CBCL Paper, p. 182 (1999)
5. Furey, T.S., Cristianini, N., Duffy, N., Bednarski, D.W., Schummer, M., Haussler, D.: Support vector machine classification and validation of cancer tissue samples using microarray expression data. Bioinformatics **16**(10), 906–914 (2000)
6. Brown, M.P.S., Grundy, W.N., Lin, D., Cristianini, N., Sugnet, C.W., Furey, T.S., Ares, M., Haussler, D.: Knowledge-based analysis of microarray gene expression data by using support vector machines. Proc. National Acad. Sci. **97**(1), 262–267 (2000)
7. Chawla, N.V., Bowyer, K.W., Hall, L.O., Kegelmeyer, W.P.: Stability of feature selection algorithms: a study on high-dimensional spaces. J. Artif. Intell. Res. **16**, 321–357 (2002)
8. Bolón-Canedo, V., Sánchez-Maroño, N., Alonso-Betanzos, A., Benítez, J.M., Herrera, F.: A review of microarray datasets and applied feature selection methods. Inf. Sci. **282**, 111–135 (2014)

Chapter 9
Software Tools

Abstract This chapter provides the users with a review of some popular software tools that can help in the design of their ensembles for feature selection. There is an important number of feature selection and ensemble learning methods already implemented and available in different platforms, so it is useful to know them before coding our own ensembles. Section 9.1 comments on the methods available in different popular software tools, such as Matlab, Weka, R, scikit-learn, or more recent and sophisticated platforms for parallel learning. Then, Sect. 9.2 gives some examples of code in Matlab.

If our goal is to build an ensemble for feature selection, it is obvious that we need to implement the feature selection algorithms involved in our design. We can implement our feature selection algorithm from scratch or use an already implemented and available algorithm. There are plenty of feature selection methods available in popular frameworks. Moreover, these frameworks often offer already available methods for distributing and combining the data in an ensemble scheme. Remember that an ensemble for feature selection can consist of applying a feature selection followed by a classification algorithm, so this scheme can benefit from the general tools for ensemble learning. And, although not so common, there are some platforms that provide implementations for ensembles for feature selection.

In the following, we will describe the most famous tools that offer frameworks for feature selection and/or ensemble learning methods, which can help us design new ensembles for feature selection. Moreover, we provide some code examples.

9.1 Popular Software Tools

9.1.1 Matlab

Matlab [1] is a numerical computing environment, well known and widely used by scientific researchers. It was developed by MathWorks in 1984 and its name comes from *Matrix Laboratory*. Matlab allows matrix manipulations, plotting of functions

© Springer International Publishing AG, part of Springer Nature 2018

V. Bolón-Canedo and A. Alonso-Betanzos, *Recent Advances in Ensembles for Feature Selection*, Intelligent Systems Reference Library 147, https://doi.org/10.1007/978-3-319-90080-3_9

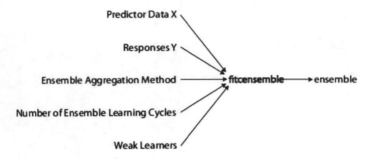

Fig. 9.1 Information you need to create an ensemble using Matlab

and data, implementation of algorithms, creation of user interfaces, and interfacing with programs written in other languages, including C, C++, Java, and Fortran.

Matlab has a set of additional toolboxes, devoted to specific problems. In particular, in the *Statistics and Machine Learning* toolbox it is possible to find the following methods for feature selection:

- `fscnca`: Feature selection using neighborhood component analysis for classification.
- `fsrnca`: Feature selection using neighborhood component analysis for regression.
- `sequentialfs`: Sequential feature selection.
- `relieff`: Importance of attributes (predictors) using ReliefF algorithm.

In the same toolbox, it is also possible to find a framework for ensemble learning. It provides a method for classification, `fitcensemble`, and for regression, `fitrensemble`. It allows the user to control parameters such as the aggregation method, the number of ensemble learning cycles and the weak learners (see Fig. 9.1).

Furthermore, Matlab provides the function `predictorImportance` in the *Statistics and Machine Learning* toolbox which, used together with an ensemble, computes estimates of predictor importance by summing these estimates over all weak learners in the ensemble, where a higher value means a more important feature.

9.1.2   Weka

Weka (Waikato Environment for Knowledge Analysis) [2] is a collection of machine learning algorithms for data mining tasks. The algorithms can be either applied directly to a dataset or called from your own Java code. Weka contains tools for data preprocessing, classification, regression, clustering, association rules, feature selection and visualization. It is also well suited for developing new machine learning schemes.

It has a wide suite of feature selection algorithms available, as described below:

- `CfsSubsetEval`: Evaluates the worth of a subset of attributes by considering the individual predictive ability of each feature along with the degree of redundancy between them.
- `ChiSquaredAttributeEval`: Evaluates the worth of an attribute by computing the value of the chi-squared statistic with respect to the class.
- `ClassifierSubsetEval`: Evaluates attribute subsets on training data or a separate hold out testing set.
- `ConsistencySubsetEval`: Evaluates the worth of a subset of attributes by the level of consistency in the class values when the training instances are projected onto the subset of attributes.
- `CostSensitiveAttributeEval`: A meta subset evaluator that makes its base subset evaluator cost-sensitive.
- `CostSensitiveSubsetEval`: A meta subset evaluator that makes its base subset evaluator cost-sensitive.
- `FilteredAttributeEval`: Class for running an arbitrary attribute evaluator on data that has been passed through an arbitrary filter (note: filters that alter the order or number of attributes are not allowed).
- `FilteredSubsetEval`: Class for running an arbitrary subset evaluator on data that has been passed through an arbitrary filter (note: filters that alter the order or number of attributes are not allowed).
- `GainRatioAttributeEval`: Evaluates the worth of an attribute by measuring the gain ratio with respect to the class.
- `InfoGainAttributeEval`: Evaluates the worth of an attribute by measuring the information gain with respect to the class.
- `LatentSemanticAnalysis`: Performs latent semantic analysis and transformation of the data.
- `OneRAttributeEval`: Evaluates the worth of an attribute by using the OneR classifier.
- `ReliefFAttributeEval`: Evaluates the worth of an attribute by repeatedly sampling an instance and considering the value of the given attribute for the nearest instance of the same and different class.
- `SVMAttributeEval`: Evaluates the worth of an attribute by using an SVM classifier.
- `SymmetricalUncertAttributeEval`: Evaluates the worth of an attribute by measuring the symmetrical uncertainty with respect to the class.
- `WrapperSubsetEval`: Evaluates attribute sets by using a learning scheme.

It also provides several methods for ensemble learning, the most popular ones are following enumerated:

- `AdaBoostM1`: Class for boosting a nominal class classifier using the Adaboost M1 method.
- `Bagging`: Class for bagging (bootstrap aggregation) a classifier to reduce variance.

- `RandomCommittee`: Class for building an ensemble of randomizable base classifiers.
- `Stacking`: Combines several classifiers using the stacking method.
- `Vote`: Class for combining classifiers.
- `RandomForest`: An extension of bagging for constructing a forest of decision trees that can be used for classification or regression.

9.1.3   R

R is a free programming language and software environment for statistical computing and graphics. The R language is widely used among statisticians and data miners for developing statistical software and data analysis. The capabilities of R are extended through user-created packages, which allow specialized statistical techniques, graphical devices, import/export capabilities, reporting tools, etc. There are several R-packages for feature selection, but probably the most famous ones are Caret and Boruta.

- The `caret`[1] package (short for Classification And REgression Training) is a set of functions that attempt to streamline the process for creating predictive models. Among other functionality, it includes some algorithms for feature selection. In particular it provides univariate feature selection, recursive feature selection, feature selection using genetic algorithms and feature selection using Simulated Annealing.
- The `Boruta`[2] package provides an all relevant feature selection wrapper algorithm. It finds relevant features by comparing original attributes' importance with importance achievable at random, estimated using their permuted copies.

There are also several packages available for ensemble learning, some of them described below:

- The `adabag`[3] package implements Freund and Schapire's Adaboost.M1 algorithm and Breiman's Bagging algorithm using classification trees as individual classifiers.
- The `randomForest`[4] package provides implementation for the popular Random Forest algorithm consisting on a forest of decision trees.
- The `gbm`[5] package (short for Generalized Boosted Regression Models) provides an implementation of extensions to Freund and Schapire's AdaBoost algorithm and Friedman's gradient boosting machine. Includes regression methods for least

---

[1] https://CRAN.R-project.org/package=caret.
[2] https://CRAN.R-project.org/package=Boruta.
[3] https://CRAN.R-project.org/package=adabag.
[4] https://CRAN.R-project.org/package=randomForest.
[5] https://CRAN.R-project.org/package=gbm.

squares, absolute loss, t-distribution loss, quantile regression, logistic, multinomial logistic, Poisson, Cox proportional hazards partial likelihood, AdaBoost exponential loss, Huberized hinge loss, and Learning to Rank measures (LambdaMart).

It is also possible to find some works providing R packages for ensemble feature selection, such as that by Neumann et al. [3]. They propose a software called EFS (Ensemble Feature Selection) available as R-package[6] and as a web application.[7] It makes use of eight feature selection methods and combines their normalized outputs to a quantitative ensemble importance. Another example, is mRMRe,[8] an R package for parallelized mRMR ensemble feature selection. The two crucial aspects of the implementation they propose are the parallelization of the key steps of the algorithm and the use of a lazy procedure to compute only the part of the mutual information minimization (MIM) that is required during the search for the best set of features (instead of estimating the full MIM).

9.1.4 KEEL

KEEL (Knowledge Extraction based on Evolutionary Learning) [4] is an open source Java software tool that can be used for a large number of different knowledge data discovery tasks. KEEL provides a simple GUI based on data flow to design experiments with different datasets and computational intelligence algorithms (paying special attention to evolutionary algorithms) in order to assess the behavior of the methods. It contains a wide variety of classical knowledge extraction algorithms, preprocessing techniques (training set selection, feature selection, discretization, imputation methods for missing values, among others), computational intelligence based learning algorithms, hybrid models, statistical methodologies for contrasting experiments and so forth. It allows to perform a complete analysis of new computational intelligence proposals in comparison to existing ones. The feature selection algorithms included in this tool are:

- `MIFS-FS`: Mutual Information Feature Selection.
- `LVF-FS`: Las Vegas Filter.
- `Focus-FS`: FOCUS.
- `Relief-FS`: Relief.
- `LVW-FS`: Las Vegas Wrapper
- `ABB-IEP-FS`: Automatic Branch and Bound using Inconsistent Examples Pairs Measure.
- `ABB-LIU-FS`: Automatic Branch and Bound using Inconsistent Examples Measure.
- `ABB-MI-FS`: Automatic Branch and Bound using Mutual Information Measure.

---

[6]https://CRAN.R-project.org/package=EFS.

[7]http://efs.heiderlab.de.

[8]https://CRAN.R-project.org/package=mRMRe.

- `Full-IEP-FS`: Full Exploration using Inconsistent Examples Pairs Measure.
- `Full-LIU-FS`: Full Exploration (LIU).
- `Full-MI-FS`: Full Exploration using Mutual Information measure.
- `Relief-F-FS`: Relief-F.
- `LVF-IEP-FS`: Las Vegas Filter using Inconsistent Examples Pairs Measure.
- `SA-IEP-FS`: Simulated Annealing using Inconsistent Examples Pairs measure.
- `SA-LIU-FS`: Simulated Annealing using Inconsistent Examples measure.
- `SA-MI-FS`: Simulated Annealing using Mutual Information measure.
- `SBS-IEP-FS`: Sequential Backward Search using Inconsistent Examples Pairs measure.
- `SBS-LIU-FS`: Sequential Backward Search using Inconsistent Examples measure.
- `SBS-MI-FS`: Sequential Backward Search using Mutual Information measure.
- `SFS-IEP-FS`: Sequential Forward Search using Inconsistent Examples Pairs measure.
- `SFS-LIU-FS`: Sequential Forward Search using Inconsistent Examples measure.
- `SFS-MI-FS`: Sequential Forward Search using Mutual Information measure.
- `SSGA-Integer-knn-FS`: Steady-state genetic algorithm with integer coding scheme for wrapper feature selection with k-NN.
- `GGA-Binary-Inconsistency-FS`: Generational genetic algorithm with binary coding scheme for filter feature selection with the inconsistency rate.
- `GGA-FS`: Generational Genetic Algorithm for Feature Selection.

It also includes several implementations of ensembles, that are following described, as well as specific methods for ensembles for imbalanced data:

- `CVCommitteesFilter-F`: Cross-Validated Committees Filter for noise elimination.
- `C45_Binarization-C`: Multiclassifier learning approach (One-vs-One / One-vs-All) with C4.5 as baseline algorithm.
- `Ensemble-C`: Ensemble Neural Network for Classification Problems.
- `Ensemble-R`: Ensemble Neural Network for Regression Problems.
- `AdaBoost.NC-C`: Adaptive Boosting Negative Correlation Learning Extension with C4.5 Decision Tree as Base Classifier.

9.1.5 RapidMiner

RapidMiner [5] is a data science software platform that provides an integrated environment for data preparation, machine learning, deep learning, text mining, and predictive analytics. It is used for business and commercial applications as well as for research, education, training, rapid prototyping, and application development and supports all steps of the machine learning process including data preparation, results visualization, model validation and optimization. It includes the following feature selection tools:

- Attribute weighting: More than 30 weighting schemes measuring the influence of attributes and forming base or weight-based selections (filter approach).
- Attribute selection: Removal of attributes unrelated to target based on a chi-square or correlation-based selection criterion or on arbitrary weighting schemes like information gain, Gini index, and others.
- Automatic optimization of selections: evolutionary, forward selection, backward elimination, weight-guided, brute-force, etc.

RapidMiner also features tools for ensemble learning, including:

- Hierarchical models.
- Combination of multiple models to form a potentially stronger model.
- Vote.
- Additive regression.
- Adaboost.
- Bayesian boosting.
- Bagging.
- Stacking.
- Classification by regression.
- Meta cost for defining costs for different error types and detecting optimal models avoiding expensive errors.

It is also possible to obtain a RapidMiner plugin, called *Feature Selection Extension*,[9] which offers the Ensemble-FS operator for ensembles for feature selection. It loops several times over subsamples of the input sample. The inner feature selection operator chosen is performed each time, and the resulting attribute weights are averaged (or somewhat combined). Then, the robustness of the feature selection can be estimated by calculating the Jaccard-Index for the different subsets of selected features.

9.1.6 Scikit-Learn

Scikit-learn [6] is a free software machine learning library for the Python programming language. It features various classification, regression and clustering algorithms including support vector machines, random forests, gradient boosting, k-means and DBSCAN, and is designed to interoperate with the Python numerical and scientific libraries NumPy and SciPy. It includes several feature selection algorithms:

- Removing features with low variance: It is a simple baseline approach that removes all features whose variance does not meet some threshold.
- Univariate feature selection: It works by selecting the best features based on univariate statistical tests, such as chi-square or mutual information.

---

[9]https://sourceforge.net/projects/rm-featselext/.

- Recursive feature elimination: Given an external estimator that assigns weights to features (such as the coefficients of a linear model), it selects features by recursively considering smaller and smaller sets of features.
- L1-based feature selection: Since linear models penalized with the L1 norm have sparse solutions, many of their estimated coefficients are zero and, therefore, the non-zero coefficients can be selected.
- Tree-based feature selection: Tree-based estimators can be used to compute feature importances, which in turn can be used to discard irrelevant features.

Apart from these algorithms already included in scikit-learn, there are other feature selection frameworks built upon it. It is particularly interesting *scikit-feature*,[10] which is an open-source feature selection repository in Python developed at Arizona State University. It contains around 40 popular feature selection algorithms, including traditional feature selection algorithms and some structural and streaming feature selection algorithms.

As for ensemble learning, it also offers several options:

- Bagging meta-estimator.
- Forests of randomized trees, including Random Forests.
- Adaboost.
- Gradient Tree Boosting, both for classification and regression.
- Voting.

Notice that the tree-based estimators enumerated above, such as Random Forests or Extremely Randomized Trees, can be used to compute feature importances, which in turn can be used to discard irrelevant features. In particular, the relative rank (i.e. depth) of a feature used as a decision node in a tree can be used to assess the relative importance of that feature with respect to the predictability of the target variable. Features used at the top of the tree contribute to the final prediction decision of a larger fraction of the input samples. The expected fraction of the samples they contribute to can thus be used as an estimate of the relative importance of the features. By averaging those expected activity rates over several randomized trees one can reduce the variance of such an estimate and use it for feature selection.

9.1.7 Parallel Learning

Most existing feature selection methods were developed decades ago and are not expected to scale efficiently when dealing with millions (or even thousands) of features; indeed, they may even become inapplicable. This is even more important in the case of ensembles, since sometimes the design of them implies to apply several feature selection methods on the same data, thus increasing the computational time. A possible solution might be to use the so-called distributed or parallel learning

---

[10]http://featureselection.asu.edu/index.php.

paradigm, which consists of distributing the learning process across several nodes or processors and then combine the results.

Several paradigms for performing parallel learning have emerged in the last years. MapReduce [7] is one such popular programming model with an associated implementation for processing and generating large data sets with a parallel, distributed algorithm on a cluster. Hadoop, developed by Cutting and Cafarella in 2005 [8], is a set of algorithms for distributed storage and distributed processing of very large datasets on computer clusters; it is built from commodity hardware and has a processing part based on MapReduce. Developed more recently, is Apache Spark [9], a fast, general engine for large-scale data processing, popular among machine learning researchers due to its suitability for iterative procedures. Developed within the Apache Spark paradigm was MLlib [10], created as a scalable machine learning library containing algorithms. It is more focused on learning algorithms, such as SVM and naive Bayes classification, k-means clustering, etc., and it also includes a few, very simple, feature selection algorithms:

- `VectorSlicer`: it is a transformer that takes a feature vector and outputs a new feature vector with a sub-array of the original features. It simply selects the features that are indicated by the user.
- `RFormula`: it selects columns specified by an R model formula. Currently it supports a limited subset of the R operators, including ' ', '.', ':', '+', and '−'.
- `ChiSqSelector`: It uses the Chi-Squared test of independence to decide which features to choose.

Moreover it is possible to find works in the literature that accelerate more sophisticated feature selection algorithm using these platforms. For example, we have developed a distributed implementation of a generic feature selection framework using Apache Spark [11] (available on GitHub[11]). This framework includes well-known information theory-based methods such as mRMR, conditional mutual information maximization, or joint mutual information (JMI), that have been designed to be able to be integrated in the Spark MLlib library. Also, we have also proposed a Spark implementation of other popular feature selection methods such as ReliefF, SVM-RFE or CFS.[12]

Apache Flink [12] is also an open-source stream processing framework for distributed, high-performing, always-available, and accurate data streaming applications. Similarly to MLlib, it has a library for machine learning for Flink, called FlinkML. However, as for now it does not include any feature selection or ensemble learning algorithms. As happens with Spark, it is possible to find works devoted to feature selection to work in Flink.[13]

Another solution to the scalability problem is the use of graphics processing units (GPUs) to distribute and thus accelerate calculations made in feature selection algorithms. With many applications to physics simulations, signal processing, financial

---

[11]https://github.com/sramirez/spark-infotheoretic-feature-selection.

[12]http://www.lidiagroup.org/index.php/en/materials-en.html.

[13]https://github.com/sramirez/flink-infotheoretic-feature-selection.

modelling, neural networks, and countless other fields, parallel algorithms running on GPUs often achieve up to 100x speedup over similar CPU algorithms. In a previous work, we have redesigned the popular mRMR method to take advantage of GPU capabilities [13], showing outstanding results (available on GitHub[14]).

Regarding ensemble learning, MLlib includes methods such as Random Forest and Gradient-boosted trees.

9.2   Code Examples

In this section, we present some simple examples to demonstrate the adequacy of using feature selection and, in particular, of using an ensemble of feature selection. The examples were coded in Matlab (see Sect. 9.1.1). For these experiments, we used a reduced version of the popular MNIST dataset, which can be downloaded here.[15] Remember that it is possible that you need additional Matlab toolboxes.

9.2.1   Example: Building an Ensemble of Trees

In this case, we are not using feature selection. We use a subset of 2000 samples from MNIST dataset, and we compare the performance of applying a single tree with the performance of using an ensemble of 15 trees. For each tree of the ensemble, we took bootstrap samples, containing 500 samples and 200 features. After executing the code, we obtain that the accuracy of the single tree is 0.5130 while the performance of the ensemble is 0.7490, so the adequacy of using an ensemble remains demonstrated.

```
1
2   %------------------------------------------------------------------%
3   % Needs statistical toolbox
4
5   clear all
6   close all
7
8
9   % Load the data
10   load MNIST2000
11
12   % Split in training and testing
13   training_data = data(1:1000,:);
14   training_labels = labels(1:1000);
15   test_data = data(1001:2000,:);
16   test_labels = labels(1001:2000,:);
17
```

---

[14]https://github.com/sramirez/fast-mRMR.

[15]http://lidiagroup.org/index.php/en/materials-en.html.

```
18
19   % Build one decision tree
20   t = classregtree(training_data,training_labels);
21   assigned_test_labels = eval(t,test_data);
22   single_accuracy = mean(test_labels == assigned_test_labels);
23   fprintf('Accuracy of a single tree = %.4f\n', single_accuracy);
24
25
26   % Build the ensemble
27   rng(2018)
28   L = 15; % Size of ensemble
29   N = 500; % Number of samples to subsample
30   M = 200; % Number of features to subsample
31   assigned_individual_labels = zeros(1000,L);
32
33   for i = 1:L
34   rp1 = randi(N,1000,1);
35   rp2 = randperm(size(training_data,2),M);
36   tr = training_data(rp1,rp2); trl = labels(rp1);
37   t = classregtree(tr,trl);
38   assigned_individual_labels(:,i) = eval(t,test_data(:,rp2));
39   end
40
41   % Find the ensemble labels
42   assigned_ensemble_labels = mode(assigned_individual_labels,2);
43   ens_accuracy = mean(test_labels == assigned_ensemble_labels);
44   fprintf('Accuracy of the ensemble  = %.4f\n', ens_accuracy);
45
46   %----------------------------------------------------------------%
```

9.2.2   Example: Adding Feature Selection to Our Ensemble of Trees

Now, we add a feature selection pre-processing step to both the single tree and the ensemble of trees. In this case, we have chosen the ReliefF filter, already available in Matlab. In this case, we need to add an extra parameter to decide the number of features we want to keep (we chose 500 when using the whole dataset and 100 for the trees forming the ensemble). After executing the code, we obtain that the accuracy of the single tree is now improved to 0.5230 and the performance of the ensemble is now 0.7530, so we can see the benefits of including feature selection in the design of our ensemble.

```
1    %----------------------------------------------------------------%
2    % Needs statistical toolbox
3
4    clear all
5    close all
6
7    % Load the data
8    load MNIST2000
```

```
9
10   % Split in training and testing
11   training_data = data(1:1000,:);
12   training_labels = labels(1:1000);
13   test_data = data(1001:2000,:);
14   test_labels = labels(1001:2000,:);
15
16   % Number of features to keep fo the single tree
17   F = 500;
18
19   % Build one decision tree
20   % First, apply ReliefF filter
21   [ranked,weights] = relieff(training_data,training_labels,10);
22   % We use the first selected F features
23   t = classregtree(training_data(:,ranked(1:F)),training_labels);
24   assigned_testing_labels = eval(t,test_data(:,ranked(1:F)));
25   single_accuracy = mean(test_labels == assigned_testing_labels);
26   fprintf('Accuracy of a single tree = %.4f\n', single_accuracy)
27
28
29   %Build the ensemble
30   rng(2018)
31   L = 15; % Size of ensemble
32   N = 500; % Number of objects to subsample
33   M = 200; % Number of features to subsample
34   F2 = 100; % Number of features to keep in the ensemble
35   assigned_individual_labels = zeros(1000,L);
36
37   for i = 1:L
38   rp1 = randi(N,1000,1);
39   rp2 = randperm(size(training_data,2),M);
40   tr = training_data(rp1,rp2); trl = labels(rp1);
41   [ranked,weights] = relieff(tr,trl,10); % Feature selection
42   t = classregtree(tr(:,ranked(1:F2)),trl);
43   ts = test_data(:,rp2);
44   assigned_individual_labels(:,i) = eval(t,ts(:,ranked(1:F2)));
45   end
46
47   % Find the ensemble labels
48   assigned_ensemble_labels = mode(assigned_individual_labels,2);
49   ens_accuracy = mean(test_labels == assigned_ensemble_labels);
50   fprintf('Accuracy of the ensemble  = %.4f\n', ens_accuracy);
51   %-----------------------------------------------------------------%
```

9.2.3   Example: Exploring Different Ensemble Sizes for Our Ensemble

In the previous examples, the size of the ensemble was fixed as 15. Now, we explore the effect of the ensemble size, and we also compare the performance of the ensemble

Fig. 9.2 Evolution of accuracy when changing the ensemble size

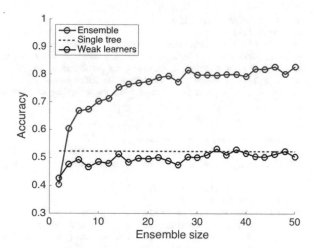

with the best performance of the individual trees that conform the ensemble (also known as weak learners). The results of executing the code is depicted in Fig. 9.2. As can be seen, the performance of the ensemble is increasing as the number of weak learners increase, and its accuracy is higher than that of the single tree and than the best one among the different weak learners in the ensemble.

```
1   %---------------------------------------------------------------%
2   % Needs statistical toolbox
3
4   clear all
5   close all
6
7
8   % Load the data
9   load MNIST2000
10
11
12  % Split in training and testing
13  training_data = data(1:1000,:);
14  training_labels = labels(1:1000);
15  test_data = data(1001:2000,:);
16  test_labels = labels(1001:2000,:);
17
18  % Number of features to keep for the single tree
19  F = 500;
20
21
22  % Build one decision tree
23  % First, apply ReliefF filter
24  [ranked,weights] = relieff(training_data,training_labels,10);
25  % We use the first selected F features
26  t = classregtree(training_data(:,ranked(1:F)),training_labels);
27  assigned_test_labels = eval(t,test_data(:,ranked(1:F)));
28  single_accuracy = mean(test_labels == assigned_test_labels);
29  fprintf('Accuracy of a single tree = %.4f\n', single_accuracy)
30
31
```

```
32   % Build the ensemble
33   rng(2018)
34   L = 50; % Maximum ensemble size to test
35   different_L = 2:2:L; % We test ensemble size increasing by 2
36   N = 500; % Number of objects to subsample
37   M = 200; % Number of features to subsample
38   F2 = 100; % Number of features to keep in the ensemble
39
40
41
42   for k = 1:length(different_L)
43   % Initialize variable when storing the labels
44   assigned_individual_labels = zeros(1000,different_L(k));
45
46   for i = 1:different_L(k)
47   rp1 = randi(N,1000,1);
48   rp2 = randperm(size(training_data,2),M);
49   tr = training_data(rp1,rp2); trl = labels(rp1);
50   [ranked,weights] = relieff(tr,trl,10); % Feature selection
51   t = classregtree(tr(:,ranked(1:F2)),trl);
52   ts = test_data(:,rp2);
53   assigned_individual_labels(:,i) = eval(t,ts(:,ranked(1:F2)));
54   % Compute the individual accuracy of each weak learner
55   ind_accuracy(i) = mean(test_labels == assigned_individual_labels(:,i));
56   end
57
58   fprintf('Max accuracy of the weak learners  = %.4f\n', max(ind_accuracy));
59   max_ind_acc(k) = max(ind_accuracy);
60   fprintf('Mean accuracy of the weak learners  = %.4f\n', mean(ind_accuracy));
61   mean_ind_acc(k) = mean(ind_accuracy);
62
63   % Find the ensemble labels
64   assigned_ensemble_labels = mode(assigned_individual_labels,2);
65   ens_accuracy(k) = mean(test_labels == assigned_ensemble_labels);
66   fprintf('Accuracy of the ensemble  = %.4f\n', ens_accuracy(k))
67   end
68
69   % Graphs
70
71   for i=1:length(different_L),
72   singleTree(i) = single_accuracy;
73   end
74
75   figure();
76   set(gca,'FontSize',18)
77   hold on;
78   plot(different_L,ens_accuracy, '-ro', 'LineWidth', 2, 'MarkerSize', 10);
79   plot(different_L,singleTree, '--b', 'LineWidth', 2);
80   plot(different_L,max_ind_acc, '-ko', 'LineWidth', 2, 'MarkerSize', 10);
81   xlabel('Ensemble size');
82   ylabel('Accuracy');
83   %ylim([0.3 1]);
84   legend('Ensemble','Single tree', 'Weak learners', 'Location', 'NorthWest');
85   hold off;
```

References

1. MATLAB. version 8.1.0.604 (R2013a). The MathWorks Inc. (2013)
2. Hall, M., Frank, E., Holmes, G., Pfahringer, B., Reutemann, P., Witten, I.H.: The Weka data mining software: an update. ACM SIGKDD Explor. Newsl. **11**(1), 10–18 (2009)
3. Neumann, U., Genze, N., Heider, D.: EFS: an ensemble feature selection tool implemented as R-package and web-application. BioData Min. **10**(1), 21 (2017)
4. Alcalá-Fernández, J., Fernández, A., Luengo, J., Derrac, J., García, S., Sánchez, L., Herrera, F.: Keel data-mining software tool: data set repository, integration of algorithms and experimental analysis framework. J. Multi.-Valued Log. Soft Comput. **17**(2–3), 255–287 (2011)
5. Hofmann, M., Klinkenberg, R.: RapidMiner: Data Mining Use Cases and Business Analytics Applications. CRC Press, Boca Raton (2013)
6. Pedregosa, F., Varoquaux, G., Gramfort, A., Michel, V., Thirion, B., Grisel, O., Blondel, M., Prettenhofer, P., Weiss, R., Dubourg, V., Vanderplas, J., Passos, A., Cournapeau, D., Brucher, M., Perrot, M., Duchesnay, E.: Scikit-learn: machine learning in python. J. Mach. Learn. Res. **12**, 2825–2830 (2011)
7. Dean, J., Ghemawat, S.: MapReduce: simplified data processing on large clusters. Commun. ACM **51**(1), 107–113 (2008)
8. Apache Hadoop. http://hadoop.apache.org/
9. Apache Spark. https://spark.apache.org
10. MLib/Apache Spark. https://spark.apache.org/mllib
11. Ramirez-Gallego, S., Mouriño-Talin, S., Martinez-Rego, D., Bolon-Canedo, V., Benitez, J.M., Alonso-Betanzos, A., Herrera, F.: An information theory-based feature selection framework for big data under Apache Spark. IEEE Trans. Syst. Man Cybern. Syst. (2018). (in press)
12. Apache Flink. https://flink.apache.org/
13. Ramirez-Gallego, S., Lastra, I., Martinez-Rego, D., Bolón-Canedo, V., Benitez, J.M., Alonso-Betanzos, A., Herrera, F.: Fast-mRMR: fast minimum redundancy maximum relevance algorithm for high-dimensional big data. Int. J. Intell. Syst. **32**(2), 134–152 (2017)

Chapter 10
Emerging Challenges

Abstract This chapter reveals the new challenges that the researchers are finding in ensemble feature selection, most of them related with "Big Data" and some of its consequences, as the important rise in unsupervised learning, because unlabelled samples is the most common situation in large datasets; or the need for visualization, that is a challenge also shared between ensemble learning and feature selection. Although feature selection is a well-established preprocessing technique, during the last years it has experimented certain renaissance due to the fact that is almost mandatory for the new scenarios in which large and/or high-dimensional datasets are present. Thus, feature selection has been successfully applied lately in areas such as DNA microarray analysis, image classification, face recognition, and text classification. Ensemble feature selection is one of the new approaches to the field, in an attempt to obtain better performances and also design distributed FS schemes that allow for more effective process and higher efficiencies. This chapter outlines some of the latest challenges in the field of ensemble feature selection, aiming researchers at following the new paths that are opened for exploration. In Sect. 10.1 a brief Introduction to the need for ensemble feature selection is outlined. Section 10.2 reviews some of the fields in which feature selection, and more specifically feature selection ensembles have been used. To end the chapter, Sect. 10.3 enumerates some of the challenges that lie ahead for feature selection, and thus for the use of ensembles in this preprocessing step.

This chapter is devoted to review briefly some of the fields in which feature selection, in general, and ensembles in particular, have contributed to improve performance. Ensemble feature selection has been applied successfully in several fields, such as microarray analysis, image classification, face recognition or text classification. In this chapter, some of these contributions are briefly described. The chapter ends with a short list of some of the current challenges in the field of feature selection, some of which perhaps can be inspiring for researchers in developing new ensemble feature selection approaches.

Part of the content of this chapter was previously published in *Knowledge-Based Systems* (https://doi.org/10.1016/j.knosys.2015.05.014.

© Springer International Publishing AG, part of Springer Nature 2018
V. Bolón-Canedo and A. Alonso-Betanzos, *Recent Advances in Ensembles for Feature Selection*, Intelligent Systems Reference Library 147, https://doi.org/10.1007/978-3-319-90080-3_10

10.1   Introduction

Feature selection is one of the most used preprocessing techniques, almost mandatory in the present scenarios of Big Data. Ongoing advances in computer-based technologies and sensorization have enabled researchers and engineers to collect data at an increasingly fast pace. In addition, data is generated in many different formats (text, multimedia, etc.) and from many different sources (systems, sensors, mobile devices, etc.). Big Data —large volumes and ultrahigh dimensionality— is now a recurring reality in most machine learning application fields, such as text mining and information retrieval [1]. To be able to extract useful information from all these data, we require new analysis and processing tools. Weinberger et al. [2], for instance, conducted a study of a collaborative email-spam filtering task with 16 trillion unique features, whereas the study by Tan et al. [1] was based on a wide range of synthetic and real-world datasets of tens of million data points with $\mathcal{O}(10^{14})$ features. The growing size of datasets raises an interesting challenge for the research community; to cite Donoho et al. [3] "our task is to find a needle in a haystack, teasing the relevant information out of a vast pile of glut".

To address the challenge of analyzing these data, feature selection becomes an imperative preprocessing step that needs to be adapted and improved to be able to handle high-dimensional data. Between the dawn of time up to 2003 humanity generated a total of 5 exabytes of data and by 2008 this figure had tripled to 14.7 exabytes. Nowadays 5 exabytes of data is produced every 2 days —and the pace of production continues to rise. Because the volume, velocity, variety and complexity of datasets is continuously increasing, machine learning techniques have become indispensable in order to extract useful information from huge amounts of otherwise meaningless data. One machine learning technique is feature selection (FS), whereby attributes that allow a problem to be clearly defined are selected, while irrelevant or redundant data are ignored. Feature selection methods have traditionally been categorized as filter methods, wrapper methods or embedded methods [4], although new approaches that combine existing methods or based on other machine learning techniques are continuously appearing to deal with the challenges of today's datasets.

Robustness or stability of feature selection techniques is a topic of recent interest, and is an important issue when selected feature subsets are subsequently analyzed by domain experts to gain more insight into the problem modelled [5, 6]. Thus, in the last few years, feature selection has been successfully applied in different scenarios involving huge volumes of data, such as DNA microarray analysis, image classification, face recognition, text classification, etc. One of the approaches that can be used for obtaining more accurate, robust and stable results in feature selection is using ensembles. Some of these ensemble approaches show great promise for high-dimensional domains with small sample sizes, and provide more robust feature subsets than a single feature selection technique [7, 8].

In general, available feature selection methods have each their merits and disadvantages. In this actual context of Big Data, their computational complexity is an important issue to take into account [9]. Nowadays, however, this factor plays a crucial role in big data problems [10]. In general, univariate methods have an important scalability advantage, but at the cost of ignoring feature dependencies and degrading classification performance. In contrast, multivariate techniques improve classification performance, but their computational burden often means that they cannot be applied to Big Data. Studies on scalability (the behavior of the feature selection methods for increasingly larger sizes of training sets) is scarce in the scientific literature. It is evident that feature selection researchers need to adapt existing methods or propose new ones in order to cope with the challenges posed by the explosion of Big Data (discussed in Sect. 10.3). Also, it is clear that the evaluation of the FS methods should be based not only on accuracy but also on execution time and stability. In order to do that, new evaluation measures are to be proposed and tested [10].

Another important issue that feature selection ensembles provide easily is diversity among the problem solvers. Then, the mechanisms to integrate the different perspectives and knowledge obtained by the individual methods becomes crucial [11]. Finally, when the feature selection methods employed are rankers, most "classical" ensemble approaches retain fixed percentages of features [9, 12]. More recent approaches have attempted to derive general automatic thresholds [13–15]. In a society that needs to deal with vast quantities of data and features in all kinds of disciplines, there is an urgent need for solutions to the indispensable issue of feature selection, some of which can be confronted using an ensemble approach. To understand the challenges that researchers face, in the next sections we will describe the most recent works in the field of feature selection, and later on enumerate the challenges that we are to face in the very near future.

10.2 Recent Contributions in Feature Selection

Several works have reviewed the most widely used feature selection methods over the last years [9]. Molina et al. [16] assessed the performance of fundamental feature selection algorithms in a controlled scenario, taking into account dataset relevance, irrelevance and redundancy. Saeys et al. [5] created a basic taxonomy of classical feature selection techniques, discussing their use in bioinformatics applications. Hua et al. [17] compared some basic feature selection methods in settings involving thousands of features, using both model-based synthetic data and real data. Brown et al. [18] presented a unifying framework for information theoretic feature selection, bringing almost two decades of research into heuristic filter criteria under a single theoretical umbrella. Finally, García et al. [19] dedicated a chapter in their data preprocessing book to a discussion of feature selection and an analysis of its main aspects and methods. New feature selection methods are constantly being developed so there is a wide suite available to researchers. Below we assess recent developments in solutions for high-dimensionality problems in areas such as clustering [20, 21], regression [22–24] and classification [25, 26].

The use of different feature types and combinations is almost standard in many of today's real applications, leading to a veritable feature explosion given rapid advances in computing and information technologies [27]. Traditionally, and due to the need of dealing with extremely high-dimensionality data, most newer feature selection approaches were filter methods. However, the last few years have come with an increasing pace in the appearance of embedded methods, as that they allow for simultaneous feature selection and classification [28–30]. Wrapper methods have received less attention, due to the heavy computational burden and the high risk of overfitting when the number of samples is insufficient. Finally, there is also a tendency to combine algorithms, either in the form of hybrid methods [31–34] or ensemble methods [6, 35–39].

Another perspective of the field can be obtained when focusing on a given application area, with researchers employing different feature selection techniques in an attempt to improve performance. In this case, some times the methodologies are highly dependent on the problem at hand. Some of the most representative applications are discussed below.

10.2.1 Applications

Ensembles for feature selection methods are currently being applied to problems in very different fields. Below we describe some of the most popular applications promoting the use of either feature selection methods or ensembles for feature selection.

10.2.1.1 Microarray Analysis

DNA microarrays are used to collect information on gene expression differences in tissue and cell samples that could be useful for disease diagnosis or for distinguishing specific types of tumours. The sample size is usually small (often less than 100 patients) but the raw data measuring the gene expression en masse may have from 6000–60,000 features. In this scenario, feature selection inevitably became an indispensable preprocessing step.

The earliest work in this field, in the 2000s [5], was dominated by the univariate paradigm [40–42], which is fast and scalable but which ignores feature dependencies. However, some attempts were also made with multivariate methods, as these can model feature dependencies, although they are slower and less scalable than univariate techniques [9]. Multivariate filter methods were used [43–46] and also more complex techniques such as wrapper and embedded methods [47–50]. A complete review of the most up-to-date feature selection methods used for microarray data can be found in [51], which indicates that many contributions since 2008 fall into the filter category, mostly based on information theory (see Fig. 10.1). The wrapper approach has largely been avoided due to the heavy computational consumption of resources and the high risk of overfitting. Although the embedded approach did not receive much attention in the infancy of microarray data classification, several proposals have emerged in recent years. Finally, it is worth noting that the recent

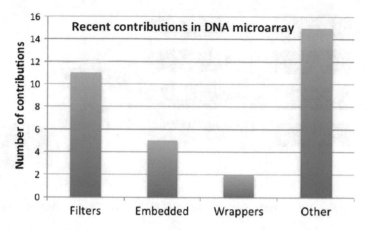

Fig. 10.1 Recent feature selection contributions to DNA microarray analysis according to the data collected in [51]

literature reveals a tendency to combine algorithms in hybrid or ensemble methods (represented as "Other" in Fig. 10.1).

Some recent works can be found in [52–54]. In the later, a primary filter step using Fisher criterion is used with the aim of reducing the initial genes, and thus the search space and time complexity. Subsequently, a wrapper approach which is based on cellular learning automata, optimized with ant colony method, is used to find the set of features which improve the classification accuracy. Finally, the selected features from the last phase are evaluated using ROC curve and the most effective while smallest feature subset is determined. In the work by [52], an approach named ensemble gene selection by grouping (EGSG), is used to select multiple gene subsets for subsequent classification. The method chooses salient gene subsets from microarray data by virtue of information theory and approximate Markov blanket, instead of employing a random selection. The experimental results show that the method improves stability over the random approach, while comparable classification performance to other gene selection methods is obtained. Finally, in [53] an ensemble of filters, only of the ranker type, is used (see Fig. 10.2). The individual rankings obtained by the four filters selected, are subsequently combined with different aggregation methods. Finally, as the filters used are all ranking methods (thus, an ordered list of all original features is returned), a threshold is necessary to obtain a practical subset of features. The approach described uses a novel proposal, that consists in using a data complexity measure [55], specifically the inverse of Fisher discriminant ratio, for establishing an automatic ranking. Using a SVM as classifier, the proposal was able to obtain the best results in several microarray datasets (see Table 10.1), that draw different scenarios regarding balance, complexity, dataset shift [51], etc.

This work has been extended later by the same authors in [56]. In this new work, the authors propose two models, that differentiate on whether thresholding was performed before or after the combination step. Different from their first work, two more

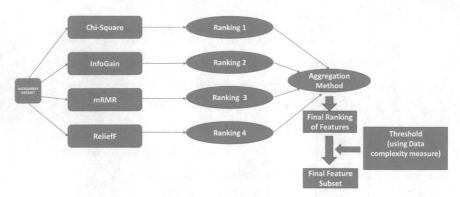

Fig. 10.2 Diagram of the proposed ensemble method for featuring feature selection in microarray datasets

Table 10.1 Binary microarray datasets employed in the experimental study

Dataset	Features	Samples		Train distribution (%)	Test distribution (%)
		Train	Test		
Colon	2000	42	20	67 – 33	60 – 40
DLBCL	4026	32	15	50 – 50	53 – 47
CNS	7129	40	20	65 – 35	65 – 35
Leukemia	7129	38	34	71 – 29	59 – 41
Lung	12533	32	149	50 – 50	90 – 10
Prostate	12600	102	34	49 – 51	26 – 74
Ovarian	15154	169	84	35 – 65	38 – 62

embedded methods were added to the ensemble, specifically SVM-RFE (Recursive Feature Elimination for Support Vector Machines) [57] and FS-P (Feature Selection-Perceptron) [58]. Besides, the combination methods selected were different regarding if ranks or subsets were to be combined, and also finally several different automatic thresholds, based all on data complexity measures and combinations of those, were tested. The experimentation was carried out over synthetic, standard real data sets (those in which the number of features is much lower than the number of samples), and also microarray datasets. The conclusion regarding automatic thresholding for microarrays is that these type of thresholds obtain much better results than the fixed percentages of features retained (25%, 50%, etc.), and that the ensemble that performs thresholding prior to combination obtains, in general, better performance results.

10.2.1.2   Image Classification

Image classification has become a popular research field given the demand for efficient ways to classify images into categories. The numerical properties of image features are usually analyzed to determine to which category they belong. With recent advances in image capture and storage and Internet technologies, a vast amount of image data has become available to the public, from smartphone photo collections to websites and even video databases. Since image processing usually requires a large amount of computer memory and power, feature selection can help in reducing the number of features needed in order to be able to correctly classify the image [12].

Although the explosion of data has evidenced the adequacy of feature selection techniques to deal with millions of images, a need to know precisely which features to extract from each pixel arose decades ago. A common problem in this field is that the literature refers to many models for extracting textural features from a given image, such as Markov random fields and co-occurrence features. However, as Ohanian and Dubes pointed out [59], there is no universally best subset of features. For this reason, the feature selection task has to be specific to each problem in order to decide which type of feature to use. Another task of interest derived from the use of the adequate features is reducing the computational time necessary to extract them. When the number of features extracted and processed is reduced, the time required is also reduced in consonance, and this can usually be achieved with minimum performance degradation, as in [60].

The use of ensembles for classification after a feature selection step is relatively common [61–65]. However, only recently feature extraction (not feature selection) ensembles have been applied to images by means of deep networks, as in [66, 67]. Thus, feature selection ensembles for image classification seems to be, at this moment, an open line of research.

10.2.1.3   Face Recognition

Identifying a human face is a complex visual recognition problem. In the last few decades, face recognition has become one of the most active research fields due to its numerous commercial and legal applications. A common application is to identify or verify a person from a digital image or a video-sourced frame by comparing selected facial features from the image with features in a facial database. An important issue in this field is to determine which image features are the most informative for recognition purposes. Unfortunately, this is no trivial task since great redundancy exists in object images; moreover, facial databases contain a large number of features but a reduced number of samples. Feature selection algorithms for face recognition have recently been suggested as a way of solving these issues.

The filter method of feature selection is a common choice, mainly due to its low computational cost compared to the wrapper or embedded methods. Yang et al. [68] presented a method based on the physical meaning of the generalized Fisher criterion in order to choose the most discriminative features for recognition. Lu et al.

[69] proposed a novel method for choosing a subset of original features containing the most essential information; called principal feature analysis (PFA), it is similar to principal component analysis (PCA) methods. This latter is employed in [70] for extracting features for a fuzzy logic ensemble system. Matos et al. [71] introduced a face recognition method based on discrete cosine transform (DCT) coefficient selection. More recently, Lee et al. [72] introduced a new color face recognition method that uses sequential floating forward search (SFFS) to obtain a set of optimal color components for recognition purposes. Other authors as [73] have tried classification ensembles after extracting low and high resolution features from images. It is also worth noting that several proposed methods based on evolutionary computation techniques have been demonstrated to be successful in this field [74–77].

Regarding the use of feature selection ensembles for face recognition, a competition, named "Ensemble Feature Selection in Face Recognition" was launched in 2012 in the International Conference for Machine Learning. The idea was related with the need of recognizing faces automatically for security reasons, and thus the organizers provided human and avatar faces, and applied an ensemble of three well-known filters for feature selection, ReliefF, Fisher Score and Chi Square, to select a small number of features in the images (around 1%) [78], obtaining very high accurate results, much better than using higher numbers of features. However, an open line of research is to devise new feature selection ensembles that may further improve computational efficiency and learning performance. Other authors, as in Mallipedi et col. [79] designed an ensemble based approach for face recognition in which feature extraction is used, and different subsets of PCA features are obtained by maximizing the distance between a subset of classes (that overlap and are obtained by bagging) of the training data instead of whole classes. Each subset of the PCA features obtained is used for face recognition and all the outputs are combined by a simple majority voting. Finally, in [80] a novel system for face recognition based on ensembles of classifiers that combine different preprocessing techniques, that in turn vary a set of feature extraction parameters. Due to the fact that Deep learning methods had significantly outperformed previous systems based on low level features in face recognition, the authors also tested their proposal using a set of features obtained from the internal representation of a Convolutional Neural Network (CNN) trained for the face recognition problem. Not only the first approach, based on hand-crafted features, obtained much better results that the state-of-the-art approaches, but the ensemble based on the fusion of learned (from CNN) and handcrafted features improves performance even further. Thus, ensemble feature selection for face recognition appears to be an interesting line of open research.

10.2.1.4   Text Classification

The goal of text classification is to categorize documents into a fixed number of predefined categories or labels. This problem has become particularly relevant to Internet applications for spam detection and shopping and auction websites. Each unique word in a document is considered a feature. However, since this implies far more input features than examples (usually by more than an order of magnitude), it is necessary to select a fraction of the vocabulary and so allow the learning algorithm to reduce computational, storage and/or bandwidth requirements.

A preprocessing stage is usually applied prior to feature selection to eliminate rare words and to merge word forms such as plurals and verb conjugations into the same term. There are several approaches to representing the feature values, for instance, a Boolean value to indicate if a word is present or absent or including the count of word occurrences. Even after this preprocessing step, the number of possible words in a document may still be high, so feature selection is paramount. A number of techniques have been developed and applied to this problem in recent years. Forman [81] proposed a novel feature selection metric, called bi-normal separation (BNS), which is a useful heuristic for increased scalability when used with wrapper techniques of text classification. Kim et al. [82] applied several novel feature selection methods to clustered data, while Dasgupta et al. [83] proposed an unsupervised feature selection strategy that theoretically guarantees the generalization power of the resulting classification function with respect to the classification function based on all the features. Forman [84] reviewed a series of filters applied to binary, multiclass and hierarchical text classification problems, focusing especially on scalability. Uguz [33] subsequently proposed a two-stage feature selection method for text categorization using InfoGain, PCA and genetic algorithms, obtaining high categorization effectiveness for two classical benchmark datasets. Shang et al. [85] recently proposed a novel metric called global information gain (GIG) that avoids redundancy naturally and also introduced an efficient feature selection method called maximizing global information gain (MGIG), which has proved to be effective for feature selection in the text domain. More recently, Baccianella et al. [86] presented six novel feature selection methods specifically devised for ordinal text classification. Another very interested approach has been followed in [87], in which the goal is to provide feedback to instructors concerning the results obtained by their student in an open-answer assessment. The only standard assessment that teachers generally may have in automatic assessment of open-response assignments is the qualification obtained by the students. In [87], the aim is to analyze the use that the students make of the corpus of words in their answers to the assessments, relating them to the qualifications obtained in the peer assessment process. The idea behind the methodology is to be able to obtain clusters of words that are used by the best and the worst clusters of responses, thus providing the instructor with a representation of what has been learned by the students, giving the former the opportunity to reshape materials that can guide the students to better achieve the learning goal. For that, the approach uses a representation of the words employed by the students in a euclidean space with semantic implications, and feature selection methods are used in order to restrict the words appearing in the clusters devised. This step is critical for an adequate initialization of the peaks of a posterior Gaussian Mixture Model, that finally obtains a number of clusters (three, in this application case, although this is configurable in the method) containing the most used words in the respective assignments. The instructors are thus provided with the words of those clusters in the respective quartiles (best and worst), in an attempt to supply a representation of the concepts learned by the students, and hopefully allowing them to possibly reorient materials and explanations for a more personalized learning for the students. In this application area, also ensembles for feature selection have been applied, as in [88, 89], obtaining better results in performance than those of the individual filter methods employed in the experiments.

As can be seen, most machine learning methods can take advantage of feature selection for preprocessing purposes, since it usually improves accuracy and reduces the computational cost of pattern recognition. Our brief review has covered the more popular applications for feature selection, but the literature describes many more application areas, including intrusion detection [1, 90–93] and machinery fault diagnosis, in which mostly features are extracted and selected after obtaining the raw signal [94–100].

10.3   The Future: Challenges Ahead for Feature Selection

During the last decade, ongoing advances in computer-based technologies and sensoring have enabled researchers and engineers to collect enormous quantities of data at increasingly fast paces. In this context, in which data analysis is needed to derive information and knowledge from that data, feature selection has become an almost indispensable step. However, most feature selection methods were devised before the Big Data phenomenon, and thus ironically most of the existent methods do not scale properly [10]. Large-scale data can be found in several application fields, such as Genomics, Proteomics, Health, Internet search, social networks, finance, business sectors, meteorology, complex physics simulations, environmental research, etc. The characteristics known as the V's in Big data (volume, velocity, variety, veracity, value, validity, etc.) bring interesting challenges to Machine learning methods, among which scalability is an essential one, in order to have workable and practical algorithms that can deal adequately with the scenarios coming into scene. In [10], the scalability of state-of-the-art feature selection methods is studied, checking their performance in an artificial controlled experimental scenario, contrasting the ability of the algorithms to select the relevant features and to discard the irrelevant ones when the dimensionality increases and without permitting noise or redundancy to obstruct this process. For analyzing scalability, some evaluation measures were defined, as different aspects needed to be addressed, not only accuracy, but also stability and computational time. Also, the measures to be used are different depending if the feature selection methods were rankers or subset methods. Beside the need of knowing the scalability properties of the existent methods, new methods that take into account the new big data scenario are to be designed, and thus an important number of challenges are emerging, representing current hot spots in feature selection research.

10.3.1 Millions of Dimensions

Nowadays machine learning methods need to be able to deal with the unprecedented scale of data. Analogous to big data, the term "big dimensionality" has been coined to refer to the unprecedented number of features arriving at levels that are rendering existing machine learning methods inadequate [27].

The widely-used UCI Machine Learning Repository [101] indicates that, in the 1980s, the maximum dimensionality of data was only about 100. By the 1990s, this number had increased to more than 1500 and, by 2009, to more than 3 million. If we focus on the number of attributes of the UCI datasets, 13 have more than 5000 features and most have a samples/features ratio below 0 —a level that potentially hinders any learning process. Illustratively, Fig. 10.3 shows the number of features of the highest dimensionality datasets included in the UCI Machine Learning Repository in the last seven years. In the popular LIBSVM Database [102] the maximum dimensionality of the data was about 62000 in the 1990s, increasing to some 16 million in the 2000 s and to more than 29 million in the 2010s; analogously, 20 of the existing 92 datasets have more than 5000 features and 11 datasets have many more features than samples. Seven of the datasets included in these two repositories in the last 9 years have dimensionality in the order of millions. Apart from these generic repositories, there are others with specific high dimensionality problems, such as the aforementioned DNA microarray classification [51] and image analysis [103, 104].

In this scenario, existing state-of-the-art feature selection methods are confronted by key challenges that potentially have negative repercussions on performance. As an example, Zhai et al. [27] pointed to more than a day of computational effort by the state-of-the-art SVM-RFE and mRMR feature selectors to crunch the data for

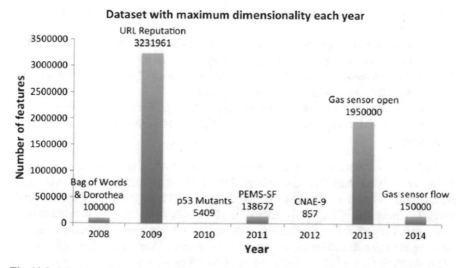

Fig. 10.3 Maximum dimensionality of the datasets included in the UCI repository [101] since 2008

a psoriasis single-nucleotide polymorphism (SNP) dataset composed of *just* half a million features.

Moreover, many state-of-the-art feature selection methods are based on algorithm designs for computing pairwise correlation. The implications when dealing with a million features are that the computer would need to handle a trillion correlations. This kind of issue poses an enormous challenge for machine learning researchers that still remains to be addressed. Some works have already attempted to use ensembles for feature selection in high dimensionality scenarios, as [105, 106].

10.3.2   Scalability

Most existing learning algorithms were developed for a much smaller dataset size, but nowadays different solutions are required for the case of small-scale versus large-scale learning problems. Small-scale learning problems are subject to the usual approximation-estimation trade-off, that is more complex in the case of large-scale learning problems, not only because of accuracy but also due to the computational complexity of the learning algorithms. Furthermore, most algorithms were designed under the assumption that the dataset would be represented as a single memory-resident table, and thus they are useless when the entire dataset does not fit in the main memory, which is the case for many datasets nowadays. Dataset size is therefore one reason for scaling up machine learning algorithms. However, there are other settings where a researcher could find the scale of a machine learning task daunting [107], for instance:

- Model and algorithm complexity: A number of high-accuracy learning algorithms either rely on complex, non-linear models, or employ computationally expensive subroutines.
- Inference time constraints: Applications that involve sensing, such as robot navigation or speech recognition, require predictions to be made in real time.
- Prediction cascades: Applications that require sequential, interdependent predictions have a highly complex joint output space.
- Model selection and parameter sweeps: Tuning learning algorithm hyper-parameters and evaluating statistical significance require multiple learning executions.

For all these reasons, scaling up learning algorithms is a trending issue. Cases in point are the workshop "PASCAL Large Scale Learning Challenge" held at the 25th International Conference on Machine learning (ICML'2008) and the "Big Learning" workshop held at the 2011 conference of the Neural Information Processing Systems Foundation (NIPS2011). Scaling up is desirable because increasing the size of the training set often increases the accuracy of algorithms [108]. In scaling up learning algorithms, the issue is not so much one of speeding up a slow algorithm as one of turning an impracticable algorithm into a practical one. Today, there is a consensus in machine learning and data mining communities that data volume presents an immediate challenge pertaining to the scalability issue [27]. The crucial point is seldom how fast you can run on a particular problem, but rather how large a problem you can deal with [109].

Scalability is defined as the impact of an increase in the size of the training set on the computational performance of an algorithm in terms of accuracy, training time and allocated memory. Thus the challenge is to find a trade-off among these criteria —in other words, to obtain "good enough" solutions as "fast" as possible and as "efficiently" as possible. As explained before, this issue becomes critical in situations in which there are temporal or spatial constraints as happens with real-time applications dealing with large datasets, unapproachable computational problems requiring learning and initial prototyping requiring rapidly implemented solutions.

Similarly to instance selection, which aims at discarding superfluous, i.e., redundant or irrelevant, samples [110], feature selection can scale machine learning algorithms by reducing input dimensionality and therefore algorithm run-time. However, when dealing with a dataset containing a huge number of both features and samples, the scalability of the feature selection method also assumes crucial importance. Since most existing feature selection techniques were designed to process small-scale data, their efficiency is likely to be downgraded, if not reduced totally, with high-dimensional data. Figure 10.4 shows run-time responses to modifications to the number of features and samples for four well-known feature selection ranker methods applied to the SD1 dataset, a synthetic dataset that simulates DNA microarray data [10].

In this scenario, feature selection researchers need to focus not only on the accuracy of the selection but also on other aspects. One such factor is stability, defined as the sensitivity of the results to training set variations. The other important factor, scalability, refers to feature selection response to an increasingly large training set. Few studies have been published regarding filter behavior in small training sets with a large number of features [18, 111–113] and even fewer on the issue of scalability [114]. What studies do exist are mainly focused on scalability in particular applications [115], modifications of existing approaches [116], combinations of instance and feature selection strategies [117] and online [118] and parallel [119] approaches. A recent paper by Tan et al. [1] describes a new adaptive feature-scaling

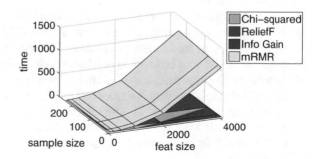

Fig. 10.4 Run-time scalability in response to modifications in the number of features and samples for four feature selection methods applied to the SD1 dataset

method —applied to several synthetic and real big datasets; based on group feature selection and multiple kernel learning, it enables scalability to big data scenarios.

Broadly speaking, although most classical univariate feature selection approaches (with each feature considered separately) have an important advantage in terms of scalability, they ignore feature dependencies and thus potentially perform less well than other feature selection techniques. Multivariate techniques, in contrast, may improve performance, but at the cost of reduced scalability [7].

Very recently, other authors [120] have presented a novel algorithm (based on the Hilbert-Schmidt independence criterion and Singular Value Decomposition) for feature selection from gene expression data, although the method can be applied to any type of problem and variables. As the algorithm does not require the whole dataset to be stored in memory, it can be scaled easily to large datasets massively distributed.

The scalability of a feature selection method is thus crucial and deserves more attention from the scientific community. One of the solutions commonly adopted to deal with the scalability issue is to distribute the data into several processors, discussed in the following section.

10.3.3   Distributed Feature Selection

As mentioned above, feature selection has been applied traditionally in a centralized manner, i.e., a single learning model is used to solve a given problem. However, since nowadays distributed data scenarios are quite common, feature selection can take advantage of processing multiple subsets in sequence or concurrently. There are several ways to distribute a feature selection task [121] (note: real-time processing will be discussed in Sect. 10.3.4):

1. The data is in one very large dataset. The data can be distributed on several processors, an identical feature selection algorithm can be run on each and the results combined.
2. The data may be in different datasets in diverse locations (e.g., in different parts of a company or even in different cooperating organizations). As for the previous case, an identical feature selection algorithm can be run on each and the results might be combined.
3. Large volumes of data may be arriving in a continuous infinite stream in real time. If the data is all streaming into a single processor, different parts can be processed by different processors acting in parallel. If the data is streaming into different processors, they can be handled as in the previous case above.
4. The dataset is not particularly large but different feature selection methods need to be applied to learn unseen instances and combine results (by some kind of voting system) [6, 8, 35, 56, 122]. The whole dataset may be in a single processor, accessed by identical or different feature selection methods that access all or part of the data. This last approach, feature selection ensemble, is the subject of this book.

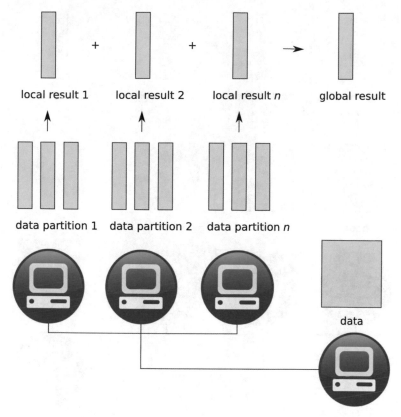

Fig. 10.5 Partitioned feature selection scenarios. The data may be in different locations, or even if the data is in one very large dataset, it might be distributed in several processors

Figures 10.5 and 10.6 show different partitioned feature selection scenarios. In the first one, Fig. 10.5 the situation described in the second case, that is, the original data is distributed between several processors and local results are combined in a final result, is represented. The second scheme, Fig. 10.6, represents the situation described in the last case above, that is, the data is replicated on different processors, local results are obtained as a consequence of applying different feature selection methods and, again, local results are combined into a global result.

As mentioned, most existing feature selection methods are not expected to scale efficiently when dealing with millions of features; indeed, they may even become inapplicable. A possible solution might be to distribute the data, run feature selection on each partition and then combine the results. The two main approaches to partitioned data distribution are by feature (vertically) or by sample (horizontally). Distributed learning has been used to scale up datasets that are too large for batch learning by samples [15, 123, 124]. While distributed learning is not common, there have been some developments regarding data distribution by features [13, 125]. One

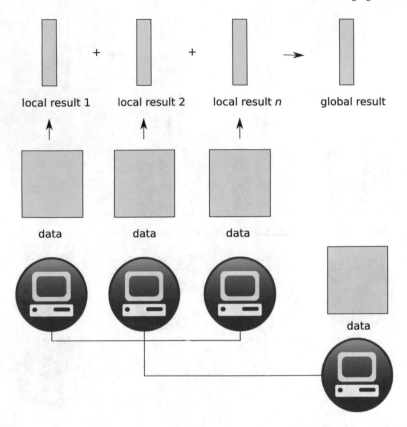

Fig. 10.6 Partitioned feature selection scenarios, the ensemble paradigm: The dataset might be in one or several locations, and one or several feature selections could be applied

proposal is a distributed method where data partitioning is both vertical and horizontal [126]. Another is a distributed parallel feature selection method that can read data in distributed form and perform parallel feature selection in symmetric multiprocessing mode via multithreading and massively parallel processing [119]. However, when dealing with big dimensionality datasets, researchers, of necessity, have to partition by features. In the case of DNA microarray data, the small sample size combined with big dimensionality prevents the use of horizontal partitioning. However, the previous mentioned vertical partitioning methods do not take into account some of the particularities of these datasets, such as the high redundancy among features, as is done in the methods described by Sharma et al. [127] and Bolón-Canedo et al. [4], the latter at a much lower computational cost. In [128] a distributed approach for partitioned data using both the two standard techniques above, horizontal (i.e. by samples) and vertical (i.e. by features) is described. Unlike other existing procedures to combine the partial outputs obtained from each partition of data, the algorithm proposed employs a merging process using the theoretical complexity of the feature subsets. The method obtains competitive results both in terms of runtime and classification accuracy.

Several paradigms for performing distributed learning have emerged in the last decade. MapReduce [9] is one such popular programming model with an associated implementation for processing and generating large data sets with a parallel, distributed algorithm on a cluster. Hadoop, developed by Cutting and Cafarella in 2005 [27], is a set of algorithms for distributed storage and distributed processing of very large datasets on computer clusters; it is built from commodity hardware and has a processing part based on MapReduce. Developed more recently, is Apache Spark [129], a fast, general engine for large-scale data processing, popular among machine learning researchers due to its suitability for iterative procedures. Developed within the Apache Spark paradigm was MLlib [130], created as a scalable machine learning library containing algorithms. Although it already includes a number of learning algorithms such as SVM and naive Bayes classification, k-means clustering, etc., as yet, it includes no feature selection algorithms. This poses a challenge for machine learning researchers, as well as offering an opportunity to initiate a new line of research. In fact, there have been several works that have developed versions of well-known algorithms that can run under Spark, as in [131–134].

Another open line of research is the use of graphics processing units (GPUs) to distribute and thus accelerate calculations made in feature selection algorithms. With many applications to physics simulations, signal processing, financial modelling, neural networks, and countless other fields, parallel algorithms running on GPUs often achieve up to 100x speedup over similar CPU algorithms. The challenge now is to take advantage of GPU capabilities to adapt existing state-of-the-art feature selection methods to be able to cope effectively and accurately with millions of features, as it has been described in [133], with a GPU version of the state-of-the-art and widely used mRMR algorithm.

10.3.4   Real-Time Processing

Data nowadays is being collected at an unprecedented fast pace and for its analysis to be useful, needs to be processed rapidly. Social media networks and portable devices dominate our day-to-day and we need sophisticated methods that are capable of dealing with vast amounts of data in real time, e.g., for spam detection and video/image detection [27].

Classical batch learning algorithms cannot deal with continuously flowing data streams, which require online approaches. Online learning [135], which is the process of continuously revising and refining a model by incorporating new data on-demand, has become a trending area in the last few years, because it solves important problems for processes occurring in time (e.g., a stock value given its history and other external factors). The mapping process is updated in real time and as more samples are obtained. Online learning can also be useful for extremely large-scale datasets, since a possible solution might be to learn data in a sequential fashion.

Online feature selection has not received the same attention as online learning [135]. Nonetheless, a few studies exist that describe attempts to select relevant features in a scenario in which both new samples and new features arise. Zhang et al. [136] proposed an incremental feature subset selection algorithm which, originating in the Boolean matrix technique, efficiently selects useful features for the given data objective. Nevertheless, the efficiency of the feature selection method was not tested with an incremental machine learning algorithm. Katakis et al. [137] proposed the idea of a dynamic feature space, whereby features selected from an initial collection of training documents are subsequently considered by the learner during system operation. However, features may vary over time and an initial training set is often not available in some applications. Thus, their proposal combined incremental feature selection with what they called a feature-based learning algorithm to deal with online learning in high-dimensional data streams. This same framework was applied to the special case of concept drift [138] inherent to textual data streams (i.e., the appearance of new predictive words over time). The approach however is limited to those datasets in which features have discrete values. Perkins et al. [139] described a novel and flexible approach, called grafting, which treats the selection of suitable features as an integral part of learning a predictor in a regularized learning framework. What makes grafting suitable for large problems is that it operates in an incremental iterative fashion, gradually building up a feature set while training a predictor model using gradient descent. Perkins and Theiler [140] tackled the problem of features arriving one at a time rather than being available from the outset; their approach, called online feature selection (OFS), assumes that, for whatever reason, it is not worthwhile waiting until all features have arrived before learning begins. They thus derived a "good enough" mapping function from inputs to outputs based on a subset of features seen so date. The potential of OFS in the image processing domain was demonstrated by applying it to the problem of edge detection [141]. A promising alternative method, called online streaming feature selection (OSFS), selects strongly relevant and non-redundant features [142]. In yet another approach, two novel online feature selection methods use relevance to select features on the fly; redundancy is only later taken into account, when these features come via streaming, but the number of training examples remains fixed [143]. Finally, the literature contains a number of studies referring to online feature selection and classification. One is an online learning algorithm for feature extraction and classification, implemented for impact acoustics signals to sort hazelnut kernels [144]. Another, by Levi and Ullman [145], proposed classifying images by ongoing feature selection, although their approach only uses a small subset of the training data at each stage. Yet another describes online feature selection performed based on the weights assigned to each classifier input [146].

As can be seen, online feature selection has been dealt with mostly on an individual basis, i.e., by pre-selecting features in a step independent of the online machine learning step, or by performing online feature selection without subsequent online classification. Therefore, achieving real-time analysis and prediction for high-dimensional datasets remains a challenge for computational intelligence on portable platforms. The question now is to find flexible feature selection methods capable of modifying the selected subset of features as new training samples arrive. It is also desirable for these methods to be executed in a dynamic feature space that would initially be empty but would add features as new information arrived (e.g., documents in their text categorization application). In this regard, in [147], an interesting method that covers both online feature selection and online learning is proposed. Notice that after an online feature selection process, where the set of relevant features might change over time, the classification algorithm has to be capable of updating its model according not only to new samples but also to new features, limiting the alternatives available capable of coping with both requirements. The proposal includes a re-implementation of the 2 metric [148] chosen due to its simplicity and effectiveness, as well as having some characteristics that make it inherently incremental. As this filter requires data to be discrete, k-means discretizer [149] was also adapted to make it incremental. The last step of the online pipeline proposed requires an incremental classifier. But those available in the literature are incremental in the instance space, but not in the feature space. Thus, an online training algorithm for one-layer artificial neural networks is also developed. The learning algorithm continuously adapts the input layer to those features, that remind might vary in number, selected at each time.

10.3.5 Feature Cost

Most of the new feature selection methods being developed focus more on removing irrelevant and redundant features rather than on the cost of obtaining those input features. The cost associated with a feature is related with different concepts. For example, a pattern in medical diagnostics consists of observable symptoms (such as age, sex, etc.), which have no cost, along with the results of tests, which are associated with costs and risks; as one example, invasive exploratory surgery is much more expensive and risky than a blood test [150]. Another example of feature extraction risk is given by Bahamonde et al. [151], where zoometry on living animals is necessary to evaluate the merits of beef cattle. Another different cost is that related to computational issues. In the medical imaging field, feature extraction from a medical image can be computationally costly; moreover, in the texture analysis technique known as co-occurrence features [152], the fact that the computational cost of extracting each feature varies implies different computational times. In real-time applications, the space complexity is negligible, whereas the time complexity is crucial [153]. Figure 10.7 shows some examples of feature cost.[1]

---

[1]Sources: "IBM Blue Gene P supercomputer" by Argonne National Laboratory's Flickr page - originally posted to Flickr as Blue Gene / PFrom Argonne National Laboratory. Uploaded using F2ComButton. Licensed under CC BY-SA 2.0 via Wikimedia Commons - http://commons.wikimedia.org/wiki/File:IBM_Blue_Gene_P_supercomputer.jpg#mediaviewer/File:IBM_Blue_Gene_P_supercomputer.jpg.
"Computed tomography of human brain - large" by Department of Radiology, Uppsala University Hospital. Uploaded by Mikael Haggstrom. Licensed under CC0 via Wikimedia Commons - http://commons.wikimedia.org/wiki/File:Computed_tomography_of_human_brain_-

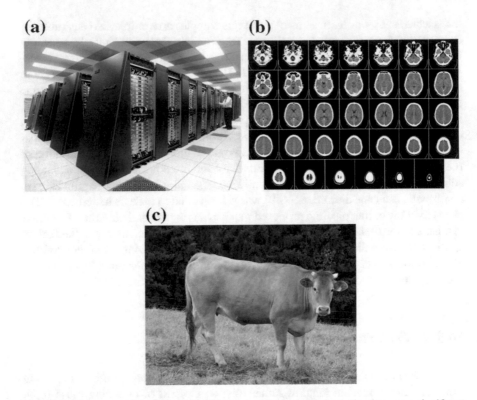

Fig. 10.7 Three examples of features with a cost. **a** Computational cost has become a significant issue in big data scenarios. **b** In medical diagnosis, the cost of a scan is not the same as the cost of a blood test. **c** Zoometry in living animals has an added cost in the form of risk

As one may notice, features with an associated cost can be found in many real-life applications. However, this has not been the focus of much attention for machine learning researchers. Most of the works have only considered the mis-classification cost, which is the penalty that is received while deciding that an object belongs to a class that it is not the real one [154].

There have been some attempts to balance the contribution of features and their cost. For instance, in classification, Friedman [155] included a regularization term to the traditional linear discriminant analysis (LDA); the left side term of their cost function evaluates error and the right side term is a regularization parameter weighted with λ, providing a framework in which different regularized solutions may appear depending on λ value. Related to feature extraction, You et al. [156] proposed a criterion to select kernel parameters based on maximizing between-class scattering and minimizing within-class scattering. A general classification framework for

_large.png#mediaviewer/File:Computed_tomography_of_human_brain_-_large.png
"Glanrind 1". Licensed under CC BY-SA 3.0 via Wikimedia Commons - http://commons. wikimedia.org/wiki/File:Glanrind_1.jpg#mediaviewer/File:Glanrind_1.jpg.

application to face recognition was proposed by Wright et al. [157] to study feature extraction and robustness to occlusion by obtaining a sparse representation. This method, instead of measuring correlation between feature and class, evaluates the representation error.

Despite the previous attempts at classification and feature extraction, there are a smaller number of works that deal with this issue in feature selection. In the early 1990s, Feddema et al. [153] developed methodologies for the automatic selection of image features by a robot. For this selection process, they employed a weighted criterion that took into account the computational cost of features, i.e., the time and space complexities of the feature extraction process. Several years later, Yang and Honavar [150] proposed a genetic algorithm to perform feature subset selection, designing the fitness function on the basis of the two criteria of neural network accuracy in classification and classification cost (defined as the cost of measuring the value of a particular feature needed for classification, the risk involved, etc.). Huang and Wang [158] also used a genetic algorithm for feature selection and parameter optimization for a support vector machine, using classification accuracy, the number of selected features and the feature cost as criteria to design the fitness function. A hybrid method for feature subset selection based on ant colony optimization and artificial neural networks has also been described [159], in which the heuristic that enables ants to select features is the inverse of the cost parameter. More recently, a new general framework was proposed that consists of adding a new term to the evaluation function of any feature selection method so that the feature cost is taken into account [160].

In Xu et al. [161], the authors examined two main components of test-time CPU cost, namely, classifier evaluation cost and feature extraction cost, and showed how to balance these costs with classification accuracy. Finally, in [162], an exponent weighted algorithm for minimal cost feature selection is proposed, in which the exponent weighted function of feature significance is constructed to increase the efficiency of the algorithm. The exponent weighted function is based on the information entropy, test cost, and a user-specified non-positive exponent.

In some other works, features are selected at testing time, providing that the whole feature subset is learned at training time from some source. In [163], an original method for providing personalized energy estimates (electricity and natural gas consumption) to prospective tenants is proposed. At training time, a cost-based forward selection algorithm selects relevant features from an establish dataset (the Residential Energy Consumption Survey), and combine low-cost features that are extractable from rental advertisements with relevant higher-cost features related to occupant behavior and home infrastructure. At test time, the aim is to make a personalized estimate for a new renter-home pair, and the algorithm dynamically orders questions for each user. These questions are based on which features inclusion would most improve the certainty of the prediction, given the information that is already known. The method show very good performance asking only a small percentage, around 20% of the total features. Other authors, as [164] propose a dynamic feature selection algorithm that automatically trades off feature cost and accuracy at the instance level. In their setting, they assume a pre-trained model using a complete set of features is given and each feature has a known cost. At test time, the aim is to dynamically select a subset of features for each instance and be able to explicitly specify the cost-accuracy trade-off. The problem is confronted as a sequential decision-making problem, employing a Markov Decision Process. This framework permits searching for an optimal purchasing policy under a reward function that combines both cost and accuracy.

Although the issue of reducing the cost associated with feature selection has received some attention in the last few years, novel feature selection methods that can deal with large-scale and real-time applications are urgently needed since computation cost must be budgeted and accounted for. The new opportunity for machine learning researchers is to match the accuracy of state-of-the-art algorithms while reducing computational cost.

10.3.6 Missing Data

Missing data is a relatively common problem in many real-world scenarios and have to be considered in practice [165]. For example, 45% of the datasets in UCI machine learning repository, which is one of the most popular collection of benchmark datasets for machine learning, contain missing values. There are several different mechanisms for which data can be missing, from problems or inadequate functioning in sensor devices, people refusing answering to some of the questions in surveys, lack of adequate equipment to measure some data in certain specific situations (for example if several hospitals are conducting an study) or simply data lost. According to the work described in [166, 167] missing data can be classified in three main types with the aim of studying testability and recoverability of the statistical analysis of the results, as missing data introduce an element of ambiguity:

1. *Missing Completely At Random (MCAR)*. This mechanism assumes that the probability of missing V_m is independent of V_m or any other variable in the study. For example, data are *MCAR* if the subjects of a survey decide to reveal their age based on flip-coins.
2. *Missing At Random (MAR)*. This mechanism assumes that for all cases Y, $P(R|Y_{obs}, Y_{miss}) = P(R|Y_{obs})$ where Y_{obs} denotes the observed component of Y and Y_{miss} the missing component. For example, data are *MAR* if men in the population are more likely to reveal their age.
3. *Missing Not At Random (MNAR)*. If the missingness mechanism are neither *MCAR* nor *MAR* are called as *MNAR*. This mechanism assumes that the value of the variable that is missing is related to the reason it is missing. For example, online shoppers rate an item with a high probability either if they love the item or if they loathe it. In other words, the probability that a shopper supplies a rating is dependent on the shopper's underlying liking [168].

There are several methods for dealing with missing data [19, 169], and several studies have been published on their influence on classification and prediction, but however its importance, very few studies have dealt with the problem of analyzing the impact that missing data has on a very commonly used and needed pre-processing step: feature selection [170]. The problem is that feature selection algorithms traditionally did not consider the missing data scenarios. In [171] the authors overcome the problem by designing an approach that can perform feature selection in high dimensional datasets without the need of previous imputation, and in [172] the authors propose a new approach, based on the concept of mutual information. The proposed procedures avoid the need for any prior imputation of the data. Other authors consider both feature selection and missing values imputation processes together, as the second can influence the results of the first. A few works tried to address this problem for classification [173], and prediction [174]. However as said above most methods address imputation or feature selection separately without considering a statistical model, in [175] the authors extend the Bayesian framework to jointly address both problems, with minimum mean square error (MMSE) estimates of missing values and by maximizing the number features correctly identified. Optimal feature selection and missing value estimation can be solved in closed form for independent Gaussian models, and fast sub-optimal methods are proposed for dependent Gaussian models. Finally, in [176], a wrapper method is devised that uses a combination of particle swarm optimization (PSO) and a classifier able to classify incomplete data (C4.5), and it has been successfully used to remove redundant/irrelevant features in incomplete data. The same authors, in [177] applied the same idea using a combination of PSO and Bagging, for the feature selection step, and again using C4.5 as classifier.

Carrying out imputation before applying feature selection is a common practice, and can introduce important bias in the data set, which effects have not been studied in depth yet. A preliminary study of the effects that different imputation strategies might have over the features, either negative or positive, as for example yielding false positives or reducing false negatives, is presented in [178]. As future work, new feature selection methods that do not require previous imputation while being robust to missing data are needed. In this context, ensembles might be a very interesting approach to test.

10.3.7 Visualization and Interpretability

In recent years, several dimensionality reduction techniques for data visualization and preprocessing have been developed. However, although the aim may be better visualization, most techniques have the limitation that the features being visualized are transformations of the original features [113, 179, 180]. Thus, when model interpretability is important, feature selection is the preferred technique for dimensionality reduction.

A model is only as good as its features, for which reason features have played and will continue to play a preponderant role in model interpretability. Users have a twofold need for interpretability and transparency in feature selection and model creation processes:

- they need more interactive model visualizations where they can change input parameters to better interact with the model and visualize future scenarios, and
- they need more interactive feature selection process where, using interactive visualizations, they are empowered to iterate through different feature subsets rather than be tied to a specific subset chosen by an algorithm.

Some recent works describe using feature selection to improve the interpretability of models obtained in different fields. One example is a method for the automatic and iterative refinement of a recommender system, in which the feature selection step selects the best characteristics of the initial model in order to automatically refine it [181]. Another is the use of feature selection to improve decision trees — representing agents simulating personnel in an organization so as to model sustainability behaviours— through an expert review of their theoretical consistency [182]. Yet another is a generative topographic mapping-based data visualization approach that estimates feature saliency simultaneously as the visualization model is trained [183]. Krause et al. [184] describe a tool in which visualization helps users develop a predictive model of their problem by allowing them to rank features (according to predefined scores), combine features and detect similarities between dimensions.

However, data is everywhere, continuously increasing, and heterogeneous. We are witnessing a form of Diogenes syndrome referring to data: organizations are collecting and storing tonnes of data, but most do not have the tools or the resources to access and generate strategic reports and insights from their data. Organizations need to gather data in a meaningful way, so as to evolve from a data-rich/knowledge-poor scenario to a data-rich/knowledge-rich scenario. The challenge is to enable user-friendly visualization of results so as to enhance interpretability. The complexity implied by big data applications also underscores the need to limit the growth in visualization complexity. Thus, even though feature selection and visualization have been dealt with in relative isolation from each other in most research to date, the visualization of data features may have an important role to play in real-world high dimensionality scenarios. However, it is also important to bear in mind that, although visualization tools are increasingly used to interpret and make complex data understandable, the quality of associated decision making is often impaired due to the fact that the tools fail to address the role played by heuristics, biases, etc. in human-computer interactive settings. Therefore, interactive tools similar to that described by Krause et al. [184] are an interesting line of research.

10.4   Summary

As can be seen throughout this book, feature selection is a much needed preprocessing step when dealing with large-scale data. It is useful for coping with scenarios with a large number of both input features and samples. However, it is especially important now that the term "Big Dimensionality" has been introduced as a consequence of the explosion of "Big Data."

The suitability of using feature selection has been demonstrated in a variety of applications that require the processing of huge amounts of data, some of which have also benefited from the ensemble paradigm for selecting adequate features. With this new scenario, there are a number of opportunities open for machine learning researchers. The need for scalable yet efficient methods is obvious, since existing feature selection methods will be inadequate for coping with this unprecedented number of features. Moreover, the society has expressed new necessities, such as distributed learning or real-time processing, where there is still an important gap that needs to be filled. However, the challenges arising in feature selection as a general field are also the challenges for the ensemble application in this regard: recent years have witnessed the creation of datasets with features numbering in the order of millions; furthermore, it seems clear that this number will only continue to increase, given the rapid advances in computing and information technologies. This new scenario offers both opportunities and challenges to machine learning researchers. There is a growing need for scalable yet efficient feature selection methods, given that existing methods are likely to prove inadequate to cope with such an unprecedented number of features. Furthermore, new needs are arising in society, such as in the areas of distributed learning and real-time processing, where an important gap that needs to be filled is developing. Beyond a shadow of doubt, the explosion in the number of features points to a number of hot spots for feature selection researchers to launch new lines of research.

References

1. Tan, M., Tsang, I.W., Wang, L.: Towards ultrahigh dimensional feature selection for big data. J. Mach. Learn. Res. **15**(1), 1371–1429 (2014)
2. Weinberger, K., Dasgupta, A., Langford, J., Smola, A., Attenberg, J.: Feature hashing for large scale multitask learning. In: Proceedings of the 26th Annual International Conference on Machine Learning, pp 1113–1120 (2009)
3. Donoho, D.L.: High-dimensional data analysis: the curses and blessings of dimensionality. In: Proceedings of the AMS Conference on Math Challenges of the 21st century, pp. 1–32 (2000)
4. Guyon, I.: Feature Extraction: Foundations and Applications. Springer, Berlin (2006)
5. Saeys, Y., Abeel, T., Van de Peer, Y.: Robust feature selection using ensemble feature selection techniques. Springer. In: Machine learning and knowledge discovery in databases (2008)
6. Seijo-Pardo, B., Porto-Díaz, I., Bolón-Canedo, V., Alonso-Betanzos, A.: Ensemble feature selection: Homogeneous and heterogeneous approaches. Knowl. Based Syst. (2017). https://doi.org/10.1016/j.knosys.2016.11.017
7. Alonso-Betanzos, A., Bolón-Canedo, Y., Fernández-Francos, D., Porto-Díaz, I., Sánchez-Maroño, N.: Up-to-Date feature selection methods for scalable and efficient machine learning. In: Igelnik, B., Zurada, J.M. (eds.) Efficiency and Scalability Methods for Computational Intellect, pp. 1–26. IGI Global (2013)
8. Seijo-Pardo, B., Bolón-Canedo, V., Alonso-Betanzos, A.: Testing different ensemble configurations for feature selection. Neural Process. Lett. **46**, 857–880 (2017)
9. Bolon-Canedo, V., Sanchez-Maroño, N., Alonso-Betanzos, A.: A review of feature selection methods on synthetic data. Knowl. Inf. Syst. **34**(3), 483–519 (2013)

10. Bolón-Canedo, V., Rego-Fernández, D., Peteiro-Barral, D., Alonso-Betanzos, A., Guijarro-Berdiñas, B., Sánchez-Maroño, N.: On the scalability of feature selection methods on high-dimensional data. Knowl. Inf. Syst. (2017). https://doi.org/10.1007/s10115-017-1140-3
11. Tsymbal, A., Pechenizkiy, M., Cunningham, P.: Diversity in search strategies for ensemble feature selection. Inf. Fusion **6**(1), 83–98 (2005)
12. Bolón-Canedo, V., Sánchez-Maroño, N., Alonso-Betanzos, A.: Recent advances and emerging challenges of feature selection in the context of big data. Knowl. Based Syst. **86**, 33–45 (2015)
13. McConnell, S., Skillicorn, D.B.: Building predictors from vertically distributed data. In: Conference of the Centre for Advanced Studies on Collaborative Research, pp. 150–162. IBM Press (2004)
14. Seijo-Pardo,B., Bolón-Canedo,V., Alonso-Betanzos,A.: Using data complexity measures for thresholding in feature selection rankers. In: Proceedings of the Advances in Artificial Intelligence. 17th Conference of the Spanish Association for Artificial Intelligence, CAEPIA Lecture Notes in Artificial Intelligence, LNAI-9868, pp 121–131 (2016)
15. Tsoumakas, G., Vlahavas, I.: Distributed data mining of large classifier ensembles. In: 2nd Hellenic Conference on Artificial Intelligence, pp. 249–256 (2002)
16. Molina, L.C., Belanche, L., Nebot: A Feature selection algorithms: a survey and experimental evaluation. In: Proceedings of the IEEE International Conference on Data Mining. ICDM 2003, pp. 306–313 (2002)
17. Hua, J., Tembe, W.D., Dougherty, E.R.: Performance of feature-selection methods in the classification of high-dimension data. Pattern Recognit. **42**(3), 409–424 (2009)
18. Brown, G., Pocock, A., Zhao, M., Luján, M.: Conditional likelihood maximisation: a unifying framework for information theoretic feature selection. J. Mach. Learn. Res. **13**(1), 27–66 (2012)
19. Garcia, S., Luengo, J., Herrera, F.: Data Preprocessing in Data Mining. Springer, Berlin (2015)
20. Chen, X., Ye, Y., Xu, X., Huang, J.Z.: A feature group weighting method for subspace clustering of high-dimensional data. Pattern Recognit. **45**(1), 434–446 (2012)
21. Song, Q., Ni, J., Wang, G.: A fast clustering-based feature subset selection algorithm for high-dimensional data. IEEE Trans. Knowl. Data Eng. **25**(1), 1–14 (2013)
22. Chen, D., Cao, X., Wen, F., Sun, J.: Blessing of dimensionality: High-dimensional feature and its efficient compression for face verification. In: IEEE Conference on Computer Vision and Pattern Recognition (CVPR), pp. 3025–3032 (2013)
23. Yamada, M., Jitkrittum, W., Sigal, L., Xing, E.P., Sugiyama, M.: High-dimensional feature selection by feature-wise kernelized Lasso. Neural Comput. **26**(1), 185–207 (2014)
24. Zhao, Z., Wang, L., Liu, H., Ye, J.: On similarity preserving feature selection. IEEE Trans. Knowl. Data Eng. **25**(3), 619–632 (2013)
25. Gan, J.Q., Hasan, B.A.S., Tsui, C.S.L.: A filter-dominating hybrid sequential forward floating search method for feature subset selection in high-dimensional space. Int. J. Mach. Learn. Cybern. **5**(3), 413–423 (2014)
26. Maldonado, S., Weber, R., Famili, F.: Feature selection for high-dimensional class-imbalanced data sets using support vector machines. Inf. Sci. **286**, 228–246 (2014)
27. Zhai, Y., Ong, Y., Tsang, I.: The emerging Big Dimensionality. IEEE Comput. Intell. Mag. **9**(3), 14–26 (2014)
28. Jawanpuria, P., Varma, M., Nath, S.: On p-norm path following in multiple kernel learning for non-linear feature selection. In: Proceedings of the 31st International Conference on Machine Learning (ICML-14), pp 118–126 (2014)
29. Maldonado, S., Weber, R., Basak, J.: Simultaneous feature selection and classification using kernel-penalized support vector machines. Inf. Sci. **181**(1), 115–128 (2011)
30. Zakharov, R., Dupont, P.: Stable Lasso for high-dimensional feature selection through proximal optimization. In: Regularization and Optimization and Kernel Methods and Support Vector Machines: Theory and Applications, Brussels and Belgium (2013)
31. Hsu, H., Hsieh, C., Lu, M.: Hybrid feature selection by combining filters and wrappers. Expert Syst. Appl. **38**(7), 8144–8150 (2011)
32. Lee, C., Leu, Y.: A novel hybrid feature selection method for microarray data analysis. Appl. Soft Comput. **11**(1), 208–213 (2011)

33. Uğuz, H.: A two-stage feature selection method for text categorization by using information gain, principal component analysis and genetic algorithm. Knowl. Based Syst. **2487**, 1024–1032 (2011)
34. Xie, J., Wang, C.: Using support vector machines with a novel hybrid feature selection method for diagnosis of erythemato-squamous diseases. Expert Syst. Appl. **38**(5), 5809–5815 (2011)
35. Bolón-Canedo, V., Sánchez-Maroño, N., Alonso-Betanzos, A.: Data classification using an ensemble of filters. Neurocomputing **135**, 13–20 (2014)
36. Haury, A.C., Gestraud, P., Vert, J.P.: The influence of feature selection methods on accuracy, stability and interpretability of molecular signatures. PloS one **6**(12), e28210 (2011)
37. Khakabimamaghani, S., Barzinpour, F., Gholamian, M.: Enhancing ensemble performance through feature selection and hybridization. Int. J. Inf. Process. Manag. **2**(2) (2011)
38. Yang, J., Yao, D., Zhan, X., Zhan, X.: Predicting disease risks using feature selection based on random forest and support vector machine. In: Bioinformatics Research and Applications. pp 1–11. Springer (2014)
39. Yang, F., Mao, K.Z.: Robust feature selection for microarray data based on multicriterion fusion. IEEE/ACM Trans. Comput. Biol. Bioinform. (TCBB) **8**(4), 1080–1092 (2011)
40. Dudoit, S., Fridlyand, J., Speed, T.P.: Comparison of discrimination methods for the classification of tumors using gene expression data. J. Am. Stat. Assoc. **97**(457), 77–87 (2002)
41. Lee, J.W., Lee, J.B., Park, M., Song, S.H.: An extensive comparison of recent classification tools applied to microarray data. Comput. Stat. Data Anal. **48**(4), 869–885 (2005)
42. Li, T., Zhang, C., Ogihara, M.: A comparative study of feature selection and multiclass classification methods for tissue classification based on gene expression. Bioinformatics **20**(15), 2429–2437 (2004)
43. Ding, C., Peng, H.: Minimum redundancy feature selection from microarray gene expression data. J. Bioinf. Comput. Biol. **3**(2), 185–205 (2005)
44. Gevaert, O., Smet, F., Timmerman, D., Moreau, Y., Moor, B.: Predicting the prognosis of breast cancer by integrating clinical and microarray data with Bayesian networks. Bioinformatics **22**(14), 184–190 (2006)
45. Wang, Y., Tetko, I.V., Hall, M.A., Frank, E., Facius, A., Mayer, K.F.X., Mewes, H.W.: Gene selection from microarray data for cancer classification: a machine learning approach. Comput. Biol. Chem. **29**(1), 37–46 (2005)
46. Yeung, K.Y., Bumgarner, R.E.: and others Multiclass classification of microarray data with repeated measurements: application to cancer. Genome Biol. **4**(12), 83–83 (2003)
47. Blanco, R., Larrañaga, P., Inza, I., Sierra, B.: Gene selection for cancer classification using wrapper approaches. Int. J. Pattern Recognit. Artif. Intell. **18**(8), 1373–1390 (2004)
48. Inza, I., Larrañaga, P., Blanco, R., Cerrolaza, A.J.: Filter versus wrapper gene selection approaches in DNA microarray domains. Artif. intell. Med. **31**(2), 91–103 (2004)
49. Jirapech-Umpai, T., Aitken, S.: Feature selection and classification for microarray data analysis: Evolutionary methods for identifying predictive genes. BMC Bioinform. **6819**, 148 (2005)
50. Ruiz, R., Riquelme, J.C., Aguilar-Ruiz, J.S.: Incremental wrapper-based gene selection from microarray data for cancer classification. Pattern Recognit. **39**(12), 2383–2392 (2006)
51. Bolón-Canedo, V., Sanchez-Maroño, N., Alonso-Betanzos, A., Benítez, J.M., Herrera, F.: A review of microarray datasets and applied feature selection methods. Inf. Sci. **282**, 111–135 (2014)
52. Liu, H., Liu, L., Zhang, H.: Ensemble gene selection by grouping for microarray data classification. J. Biomed. Inf. **43**(1), 81–87 (2010)
53. Seijo-Pardo, B., Bolón-Canedo, V., Alonso-Betanzos, A.: Using a feature selection ensemble on DNA microarray datasets. Proc. Eur. Symp. Artif. Neural Netw. Comput. Intell. Mach. Learn. ESANN **2016**, 277–282 (2016)
54. Sharbaf, F.V., Mosafer, S., Moattar, M.H.: A hybrid gene selection approach for microarray data classification using cellular learning automata and ant colony optimization. Genomics **107**(6), 231–238 (2016)
55. Basu, M., Ho, T.K.: Data Complexity in Pattern Recognition. Springer Science & Business Media (2006)

56. Seijo-Pardo, B., Bolón-Canedo, V., Alonso-Betanzos, A.: On developing an automatic threshold applied to feature selection ensembles. Inf. Fusion (2018). https://doi.org/10.1016/j.inffus. 2018.02.007
57. Guyon, I., Weston, J., Barnhill, S., Vapnik, V.: Gene selection for cancer classification using support vector machines. Mach. Learn. **46**, 1–3 (2002)
58. Mejía-Lavalle, M., Sucar, E., Arroyo, G.: Feature selection with a perceptron neural net. In: Proceedings of the International Workshop on Feature Selection for Data Mining, pp. 131–135 (2006)
59. Ohanian, P.P., Dubes, R.C.: Performance evaluation for four classes of textural features. Pattern Recognit. **25**(8), 819–833 (1992)
60. Remeseiro, B., olón-Canedo, V., Peteiro-Barral, D., Alonso-Betanzos, A., Guijarro-Berdiñas, B., Mosquera, A., Penedo, M.G., Sánchez-Maroño, N.: A methodology for improving tear film lipid layer classification. IEEE J. Biomed. Health Inf. **18**(4), 1485–1493 (2014)
61. Chowriappa, P., Dua, S., Acharya, U.R., Krishnan, M.M.R.: Ensemble selection for feature-based classification of diabetic maculopathy images. Comput. Biol. Med. **43**(12), 2156–2162 (2013)
62. Goh, J., Thing, V.L.L.: A hybrid evolutionary algorithm for feature and ensemble selection in image tampering detection. Int. J. Electron. Secur. Digit. Forensics **7**(1), 76–104 (2015)
63. Huda, S., Yearwood, J., Jelinek, H.F., Hassan, M.M., Fortino, G., Buckland, A.: Hybrid feature selection with ensemble classification for imbalanced healthcare data: a case study for brain tumor diagnosis. IEEE Access **4**, 9145–9154 (2016)
64. Sivapriya, T.R., Kamal ARNB, Thangaiah, P.R.J.: Ensemble merit merge feature selection for enhanced multinomial classification in alzheimers dementia. Comput. Math. Methods Med. 676129 (2015). https://doi.org/10.1155/2015/676129
65. Varol, E., Gaonkar, B., Erus, G., Schultz, R., Davatzikos, C.: Feature ranking based nested Support Vector Machine ensemble for medical image classification. In: Proceedings IEEE International Symposium on Biomedical Imaging: From Nano To Macro IEEE International Symposium on Biomedical Imaging, pp. 146–149 (2012). https://doi.org/10.1109/ISBI.2012. 6235505
66. Reeve, H.W.J., Brown, G.: Modular Autoencoders for ensemble feature extraction. J. Mach. Learn. Res. **44**, 242–259 (2015). NIPS
67. Tang, S., Pan, T.: Feature Extraction via Recurrent Random Deep Ensembles and its Application in Group-level Happiness Estimation. arXiv:1707.09871v1 [cs.CV] 24 Jul 2017 (2017)
68. Yang, J., Zhang, D., Yong, X., Yang, J.: Two-dimensional discriminant transform for face recognition. Pattern Recognit. **38**(7), 125–1129 (2005)
69. Lu, J., Zhao, T., Zhang, Y.: Feature selection based-on genetic algorithm for image annotation. Knowl. Based Syst. **21**(8), 887–891 (2008)
70. Polyakova, A., L Lipinskiy, L.: A study of fuzzy logic ensemble system performance on face recognition problem. IOP Conf. Ser. Mater. Sci. Eng. **173**(1), 012013 (2017)
71. de S Matos, F.M., Batista, L.V.: and others Face recognition using DCT coefficients selection. In: Proceedings of the 2008 ACM symposium on Applied computing, pp. 1753–1757 (2008)
72. Lee, S.H., Choi, J.Y., Plataniotis, K.N., Ro, Y.M.: Color component feature selection in feature-level fusion based color face recognition. In: Proceedings of the IEEE International Conference on Fuzzy Systems (FUZZ), pp. 1–6 (2010)
73. Su, Y., Shan, S., Chen, X., Gao, W.: Hierarchical ensemble of global and local classifiers for face recognition. IEEE Trans. Image Process. **18**(8), 1885–1896 (2009)
74. Amine, A., El Akadi, A., Rziza, M., Aboutajdine, D.: Ga-SVM and mutual information based frequency feature selection for face recognition, GSCM-LRIT, Faculty of Sciences, p. 1014. Mohammed V University, BP (2009)
75. Kanan, H.R., Faez, K.: An improved feature selection method based on ant colony optimization (ACO) evaluated on face recognition system. Appl. Math. Comput. **205**(2), 716–725 (2008)
76. Mazumdar, D., Mitra, S., Mitra, S.: Evolutionary-rough feature selection for face recognition. In: Transactions on Rough Sets XII, pp. 117–142. Springer, Berlin (2010)

77. Ramadan, R.M., Abdel-Kader, R.F.: Face recognition using particle swarm optimization-ba. selected features. Int. J. Signal Process. Image Process. Pattern Recognit. **2**(2), 51–65 (2009)

78. Alelyani, S., Liu, H.: Ensemble feature selection in face recognition: ICMLA 2012 challenge. In: Proceedings of the 11th International Conference on Machine Learning and Applications, pp. 588–591 (2012). https://doi.org/10.1109/ICMLA.2012.182

79. Mallipeddi, R., Lee, M.: Ensemble based face recognition using discriminant PCA Features. In: Proceedings of the IEEE Congress on Evolutionary Computation, pp. 1–7 (2012)

80. Lumini, A., Nanni, L., Brahnam, S.: Ensemble of texture descriptors and classifiers for face. Appl. Comput. Inf. **13**(1), 79–91 (2017)

81. Forman, G.: An extensive empirical study of feature selection metrics for text classification. J. Mach. Learn. Res. **3**, 1289–1305 (2003)

82. Kim, H., Howland, P., Park, H.: Dimension reduction in text classification with support vector machines. J. Mach. Learn. Res. 37–53 (2005)

83. Dasgupta, A., Drineas, P., Harb, B., Josifovski, V., Mahoney, M.V.: Feature selection methods for text classification. In: Proceedings of the 13th ACM SIGKDD international conference on Knowledge discovery and data mining, pp. 230–239(2007)

84. Forman, G.: Feature Selection for Text Classification. Computational methods of feature selection, pp. 257–276 (2008)

85. Shang, C., Li, M., Feng, S., Jiang, Q., Fan, J.: Feature selection via maximizing global information gain for text classification. Knowl. Based Syst. **54**, 298–309 (2013)

86. Baccianella, S., Esuli, A., Sebastiani, F.: Feature selection for ordinal text classification. Neural Comput. **26**(3), 557–591 (2014)

87. Bolón-Canedo,V, Diez, J. Luaces, O., Bahamonde, A., Alonso-Betanzos, A.: Paving the way for providing teaching feedback in automatic evaluation of open response assignments. In: Proceedings of the 2017 International Joint Conference on Neural Networks (IJCNN), CFP17-US-DVD (2017)

88. Shravankumar B., Ravi V.: Text classification using ensemble features selection and data mining techniques. In: Proceedings of the Swarm, Evolutionary, and Memetic Computing. SEMCCO 2014. Lecture notes in computer science, vol. 8947 (2015)

89. Van Landeghem, S., Abeel, T., Saeys, Y., Van de Peer, Y.: Discriminative and informative features for biomolecular text mining with ensemble feature selection. Bioinformatics **26**(18), i554–60 (2010)

90. Alazab, A., Hobbs, M., Abawajy, J., Alazab, M.: Using feature selection for intrusion detection system. In: Proceedings of the International Symposium on Communications and Information Technologies (ISCIT), pp. 296–301 (2012)

91. Balasaraswathi, V.R., Sugumaran, M., Hamid, Y.: Feature selection techniques for intrusion detection using non-bio-inspired and bio-inspired optimization algorithms. J. Commun. Inf. Netw. **2**(4), 107–119 (2017)

92. Hasan, M.A.M., Nasser, M., Ahmad, S., Molla, K.I.: Feature selection for intrusion detection using random forest. J. Inf. Secur. **7**, 129–140 (2016)

93. Zuech, R., Khoshgoftaar, T.M.: A survey on Feature Selection for Intrusion detection. In: Proceedings of the 21st ISSAT International Conference on Reliability and Quality in Design, pp. 150–155 (2015)

94. Chebel-Morello, B., Malinowski, S., Senoussi, H.: Feature selection for fault detection systems: application to the tennessee eastman process. Appl. Intell. **44**(1), 111–122 (2016)

95. Hui, K.H., Ooi, C.S., Lim, M.H., Leong, M.S., Al-Obaidi, S.M.: An improved wrapper-based feature selection method for machinery fault diagnosis. Plos One **12**(12), e0189143 (2017)

96. Islam, M.R., Islam, M.M.M., Kim, : Feature selection techniques for increasing reliability of fault diagnosis of bearings. In: Proceedings of the 9th International Conference on Electrical and Computer Engineering (ICECE), pp. 396–399 (2016)

97. Li, B., Zhang, P., Tian, H., Mi, S., Liu, D., Ren, G.: A new feature extraction and selection scheme for hybrid fault diagnosis of gearbox. Expert Syst. Appl. **38**(8), 10000–10009 (2011)

98. Li, H., Zhao, J., Zhang, X., Ni, X.: Fault diagnosis for machinery based on feature selection and probabilistic neural network. Int. J. Perform. Eng. **13**(7), 1165–1170 (2017)

9. Luo, M., Li, C., Zhang, X., Li, R., An, X.: Compound feature selection and parameter optimization of ELM for fault diagnosis of rolling element bearings. ISA Trans. **65**, 556–566 (2016)

100. Rajeswari, C., Sathiyabhama, B., Devendiran, S., Manivannan, K.: Bearing fault diagnosis using multiclass support vector machine with efficient feature selection methods international. J. Mech. Mechatron. Eng. **15**(1), 1–12 (2016)

101. K. Bache., M. Lichman.: UCI Machine Learning Repository, University of California, Irvine, School of Information and Computer Sciences (2013). http://archive.ics.uci.edu/ml

102. Chang, C.C., Lin, C.J.: LIBSVM: a library for support vector machines. ACM Trans. Intell. Syst. Technol. **2**(3), 27 (2011)

103. Cornell University VIA Databases. http://www.via.cornell.edu/databases

104. ImageNet. http://image-net.org

105. Brahim, A.B., Limam, M.: Ensemble feature selection for high dimensional data: a new method and a comparative study. Adv. Data Anal. Classif. 1–16 (2017)

106. Pes, B., Dess, N., Angioni, M.: Exploiting the ensemble paradigm for stable feature selection: a case study on high-dimensional genomic data. Inf. Fusion **35**, 132–147 (2017)

107. Bekkerman, R., Bilenko, M., Langford, J.: Scaling Up Machine Learning: Parallel and Distributed Approaches. Cambridge University Press, Cambridge (2011)

108. Catlett, J.: Megainduction: machine learning on very large databases. Ph.D. thesis, University of Sydney (1991)

109. Provost, F., Kolluri, V.: A survey of methods for scaling up inductive algorithms. Data Min. Knowl. Discov. **3**(2), 131–169 (1999)

110. Olvera-López, J.A., Carrasco-Ochoa, J.A., Martínez-Trinidad, J.F., Kittler, J.: A review of instance selection methods. Artif. Intell. Rev. **34**(2), 133–143 (2010)

111. Dernoncourt, D., Hanczar, B., Zucker, J.D.: Analysis of feature selection stability on high dimension and small sample data. Comput. Stat. Data Anal. **71**, 681–693 (2014)

112. Fahad, A., Tari, Z., Khalil, I., Habib, I., Alnuweiri, H.: Toward an efficient and scalable feature selection approach for Internet traffic classification. Comput. Netw. **57**, 2040–2057 (2013)

113. Gulgezen, G., Cataltepe, Z., Yu, L.: Stable and accurate feature selection. In: Machine Learning and Knowledge Discovery in Databases, pp. 455–468. Springer, Berlin (2009)

114. Peteiro-Barral, D., Bolón-Canedo, V., Alonso-Betanzos, A., Guijarro-Berdiñas, B., Sánchez-Maroño, N.: Scalability analysis of filter-based methods for feature selection. Adv. Smart Syst. Res. **2**(1), 21–26 (2012)

115. Luo, D., Wang, F., Sun, J., Markatou, M., Hu, J., Ebadollahi, S.: SOR: Scalable orthogonal regression for non-redundant feature selection and its healthcare applications. In: SIAM Data Mining Conference, pp. 576–587 (2012)

116. Sun, Y., Todorovic, S., Goodison, S.: A feature selection algorithm capable of handling extremely large data dimensionality. In: SIAM International Conference in Data Mining, pp. 530–540 (2008)

117. García-Pedrajas, N., de Haro-García, A., Pérez-Rodríguez, J.: A scalable memetic algorithm for simultaneous instance and feature selection. Evol. Comput. **22**(1), 1–45 (2014)

118. Hoi, S.C.H., Wang, J., Zhao, P., Jin, R.: Online feature selection for mining big data. In: 1st International Workshop on Big Data, Streams and Heterogeneous Source Mining: Algorithms, Systems, Programming Models and Applications, pp. 93–100. ACM (2012)

119. Zhao, Z., Zhang, R., Cox, J., Duling, D., Sarle, W.: Massively parallel feature selection: an approach based on variance preservation. Mach. Learn. **92**, 195–220 (2013)

120. Gangeh, Zarkoob, H., Ghodsi, A.: Fast and scalable feature selection for gene expression data using hilbert-schmidt independence criterion. IEEE/ACM Trans. Comput. Biol. Bioinform. **4**(1), 167–181 (2017)

121. Bramer, M.: Principles of Data Mining. Springer, Berlin (2007)

122. Bolón-Canedo, V., Sánchez-Maroño, N., Alonso-Betanzos, A.: An ensemble of filters and classifiers for microarray data classification. Pattern Recognit. **45**(1), 531–539 (2012)

123. Ananthanarayana, V.S., Subramanian, D.K., Murty, M.N.: Scalable, Distributed and Dynamic Mining of Association Rules. High performance computing, pp. 559–566 (2000)

124. Chan, P.K., Stolfo, S.J.: Toward parallel and distributed learning by meta-learning. In: AAAI Workshop in Knowledge Discovery in Databases, pp. 227–240 (1993)
125. Skillicorn, D.B., McConnell, S.M.: Distributed prediction from vertically partitioned data. J. Parallel Distrib. Comput. **68**(1), 16–36 (2008)
126. Banerjee, M., Chakravarty, S.: Privacy preserving feature selection for distributed data using virtual dimension. In: Proceedings of the 20th ACM International Conference on Information and Knowledge Management, pp. 2281–2284. ACM (2011)
127. Sharma, A., Imoto, S., Miyano, S.: A top-r feature selection algorithm for microarray gene expression data. IEEE/ACM Trans. Comput. Biol. Bioinform. (TCBB) **9**, 237–252 (2012)
128. Morán-Fernández, L., Bolón Canedo, V., Alonso-Betanzos, A.: Centralized vs. distributed feature selection methods based on data complexity measures. Knowl. Based Syst. **117**, 24–45 (2017)
129. Apache Spark. https://spark.apache.org
130. MlLib/Apache Spark. https://spark.apache.org/mllib
131. Eiras-Franco, C., Bolón Canedo, V., Ramos, S., González-Domínguez, J., Alonso-Betanzos, A., Touriño, J.: Multithreaded and Spark parallelization of feature selection filters. J. Comput. Sci. **17**, 609–619 (2016)
132. Palma-Mendoza, R.J., Rodriguez, D., e-Marcos, L.: Distributed ReliefF-based feature selection in Spark. Knowl. Inf. Syst. (2018). https://doi.org/10.1007/s10115-017-1145-y
133. Ramírez Gallego, S., Lastra, I., Martínez Rego, D., Bolón Canedo, V., Benítez, J.M., Herrera, F., Alonso Betanzos, A.: FastmRMR: fast minimum redundancy maximum relevance algorithm for high dimensional big data. Int. J. Intell. Syst. **32**(2), 154–152 (2017)
134. Ramírez-Gallego, S., Mouriño-Talín, H., Martínez-Rego, D., Bolón-Canedo, V., Benítez, J.M., Alonso-Betanzos, A., Herrera, F.: An information theory-based feature selection framework for big data under apache spark. IEEE Trans. Syst. Man Cybern. Syst. **99**, 1–13 (2017)
135. Fontenla-Romero, O., Guijarro-Berdiñas, B., Martínez-Rego, D., Pérez-Sánchez, B., Peteiro-Barral, D.: Online machine learning. In: Efficiency and Scalability Methods for Computational Intellect, pp. 27–54. IGI Global Eds (2013)
136. Zhang, C., Ruan, J., Tan, Y.: An incremental feature subset selection algorithm based on boolean matrix in decision system. Converg. Inf. Technol. 16–23 (2011)
137. Katakis, I., Tsoumakas, G., Vlahavas, I.: Dynamic feature space and incremental feature selection for the classification of textual data streams. In: Knowledge Discovery from Data Streams, pp. 107–116 (2006)
138. Tsymbal, A.: The problem of concept drift: definitions and related work. Comput. Sci. Dept. **106** (2004). Trinity College Dublin
139. Perkins, S., Lacker, K., Theiler, J.: Grafting: fast, incremental feature selection by gradient descent in function space. J. Mach. Learn. Res. **3**, 1333–1356 (2003)
140. Perkins, S., Theiler, J.: Online feature selection using grafting. In: International Conference on Machine Learning, pp. 592–599 (2003)
141. Glocer, K., Eads, D., Theiler, J.: Online feature selection for pixel classification. In: 22nd International Conference on Machine Learning, pp. 249–256 (2005)
142. Wu, X., Yu, K., Wang, H., Ding, W.: Online streaming feature selection. In: 27nd International Conference on Machine Learning, pp. 1159–1166 (2010)
143. Wu, X., Yu, K., Ding, W., Wang, H., Zhu, X.: Online feature selection with streaming features. IEEE Trans. Pattern Anal. Mach. Intell. **35**(5), 1178–1192 (2013)
144. Kalkan, H., Çetisli, B.: Online feature selection and classification. In: IEEE International Conference on Acoustics, Speech and Signal Processing, pp. 2124–2127 (2011)
145. Levi, D., Ullman, S.: Learning to classify by ongoing feature selection. Image Vis. Comput. **28**(4), 715–723 (2010)
146. Carvalho, V.R., Cohen, W.W.: Single-pass online learning: performance, voting schemes and online feature selection. In: 12th ACM SIGKDD International Conference on Knowledge Discovery and Data Mining, pp. 548–553 (2006)
147. Bolón Canedo, V., Fernández-Francos, D., Peteiro-Barral, D., Alonso Betanzos, A., Guijarro-Berdiñas, B., Sánchez-Maroño, N.: A unified pipeline for online feature selection and classification. Expert Syst. Appl. **55**, 532–545 (2016)

148. Liu, H., Setiono, R., Chi2: Feature selection and discretization of numeric attributes. In: Proceedings of the 7th International Conference on Tools with artificial intelligence, pp. 388–391 (1995)

149. Ventura, D., Martinez, T.: An empirical comparison of discretization methods. In: Proceedings of the 10th International Symposium on Computer and Information Sciences, pp. 443–450 (1995)

150. Yang, J., Honavar, V.: Feature subset selection using a genetic algorithm. IEEE Intell. Syst. Appl. **13**(2), 44–49 (1998)

151. Alonso-González, J., Bahamonde, A., Villa, A., Rodríguez-Castañón, A.A.: Morphological assessment of beef cattle according to carcass value. Livest. Sci. **107**(2–3), 265–273 (2007)

152. Haralick, R.M., Shanmugam, K., Dinstein, I.: Textural features for image classification. IEEE Trans. Syst. Man Cybern. **3**(6), 610–621 (1973)

153. Feddema, J.T., Lee, C.S.G., Mitchell, O.R.: Weighted selection of image features for resolved rate visual feedback control. IEEE Trans. Robot. Autom. **7**(1), 31–47 (1991)

154. Zhao, H., Min, F., Zhu, W.: Cost-sensitive feature selection of numeric data with measurement errors. J. Appl. Math. (2013)

155. Friedman, J.H.: Regularized discriminant analysis. J. Am. Stat. Assoc. **84**(405), 165–175 (1989)

156. You, D., Hamsici, O.C., Martinez, A.M.: Kernel optimization in discriminant analysis. IEEE Trans. Pattern Anal. Mach. Intell. **33**(3), 631–638 (2011)

157. Wright, J., Yang, A.Y., Ganesh, A., Sastry, S.S., Ma, Y.: Robust face recognition via sparse representation. IEEE Trans. Pattern Anal. Mach. Intell. **31**(2), 210–227 (2009)

158. Huang, C.L., Wang, C.J.: A ga-based feature selection and parameters optimization for support vector machines. Expert Syst. Appl. **31**(2), 231–240 (2006)

159. Sivagaminathan, R.K., Ramakrishnan, S.: A hybrid approach for feature subset selection using neural networks and ant colony optimization. Expert Syst. Appl. **33**(1), 49–60 (2007)

160. Bolón-Canedo, V., Porto-Díaz, I., Sánchez-Maroño, N., Alonso-Betanzos, A.: A framework for cost-based feature selection. Pattern Recognit. **47**(7), 2481–2489 (2014)

161. Xu, Z., Kusner, M.J., Weinberger, K.Q., Chen, M., Chapelle, O.: Classifier cascades and trees for minimizing feature evaluation cost. J. Mach. Learn. Res. **15**(1), 2113–2144 (2014)

162. Li, X., Zhao, H., Zhu, W.: An exponent weighted algorithm for minimal cost feature selection. Int. J. Mach. Learn. Cybern. **7**(5), 689–698 (2016)

163. Early, K., Fienberg, S., Mankoff, J.: Cost-Effective Feature Selection and Ordering for Personalized Energy Estimates, In Proceedings of the Thirtieth AAAI Conference on Artificial Intelligence for Smart Grids and Smart Buildings, Technical report WS-16-04, 2016

164. He, H., Daumé III, H., Esiner, J.: Cost-sensitive Dynamic Feature Selection. In: International Conference on Machine Learning (ICML) workshop on Inferning: Interactions between Inference and Learning (2012)

165. Schafer, J.L., Graham, J.W.: Missing data: our view of the state of the art. Psychol. Methods **7**(2), 147–177 (2002)

166. Pearl, J., Mohan, K.: Recoverability and testability of missing data: introduction and summary of results. SSRN 2343873, 2013

167. Rubin, D.B.: Inference and missing data. Biometrika **63**(3), 581–592 (1976)

168. Marlin, B., Zemel, R.S., Roweis, S.T., Slaney, M.: Recommender Systems. Missing Data Stat. Model Estim. IJCAI Proc. Int. Joint Conf. Artif. Intell. **22**(3), 2686 (2011)

169. Enders, C.K.: Applied Missing Data Analysis. Guilford Press (2010)

170. Guyon, I., Elisseeff, A.: An introduction to variable and feature selection. J. Mach. Learn. Res. **3**(3), 1157–1182 (2003)

171. Lou, Q., Obradovic, Z.: Margin-based feature selection in incomplete data. In: Proceedings of the AAAI, pp. 1040–1046 (2012)

172. Doquire, G., Verleysen, M.: Feature selection with missing data using mutual information estimators. Neurocomputing **90**, 3–11 (2012)

173. Zaffalon, M., Hutter, M.: Robust feature selection by mutual information distributions. In: Proceedings of the Eighteenth Conference on Uncertainty in Artificial Intelligence, pp. 577–584 (2002)

174. Meesad, P., Hengpraprohm, K.: Combination of knn-based feature selection and knn-based missing-value imputation of microarray data. In: Proceedings of the 3rd International Conference on Innovative Computing Information and Control, ICICIC'08., pp. 341–341 (2008)

175. Pour A.F., Dalton, L.A.: Optimal Bayesian feature selection with missing data. In: Proceedings of the 2016 IEEE Global Conference on Signal and Information Processing (GlobalSIP), pp. 35–39 (2016)

176. Tran, C.T., Zhang, M., Andreae, P., Xue, B.: Improving performance for classification with incomplete data using wrapper-based feature selection. Evolutionary Intell. **9**(3) (2016). https://doi.org/10.1007/s12065-016-0141-6

177. Tran, C.T., Zhang, M., Andreae, P., Xue, B.: Bagging and Feature Selection for Classification with Incomplete Data. In: Proceeding of the Evostar (2017)

178. Seijo-Pardo, B., Alonso-Betanzos, A., Bennett, K., Bolón-Canedo,V., Guyon, I., Saeed, M.: Analysis of imputation bias for feature selection with missing data. In: Proceedings of the 24th European symposium on Artificial Neural Networks, Computational Intelligence and Machine Learning (ESANN) (2018)

179. Bunte, K., Biehl, M., Hammer, B.: A general framework for dimensionality-reducing data visualization mapping. J. Neural Comput. **24**, 771–804 (2012)

180. Shalev-Shwartz, S., Ben-David, S.: Understanding Machine Learning: From Theory to Algorithms. Cambridge University Press, Cambridge (2014)

181. Bellogín, A., Cantador, I., Castells, P., Ortigosa, A.: Preference Learning, pp. 429–455. Springer, Berlin (2010)

182. Sánchez-Maroño, N., Alonso-Betanzos, A., Fontenla-Romero, O., Brinquis-Núñez, C., Polhil, J.G., Craig, T., Dumitru, A., García-Mira, R.: An agent-based model for simulating environmental behavior in a educational organization. Neural process. Lett. **42**, 89–118 (2015)

183. Maniyar, D.M., Nabney, I.T.: Data visualization with simultaneous feature selection. In: Proceedings IEEE Symposium on Computational Intelligence and Bioinformatics and Computational Biology, CIBCB'06, pp. 1–8 (2006)

184. Krause, J., Perer, A., Bertini, E.: INFUSE: interactive feature selection for predictive modeling of high dimensional data. IEEE Trans. Vis. Comput. Graph. **20**(12), 1614–1623 (2014)

185. Brown, G., Pocock, A., Zhao, M.J., Luján, M.: Conditional likelihood maximisation: a unifying framework for information theoretic feature selection. J. Mach. Learn. Res. **13**(1), 27–66 (2012)

186. Kuncheva, L.I., Whitaker, C.J.: Measures of diversity in classifier ensembles and their relationship with the ensemble accuracy. Mach. Learn. **51**(2), 181–207 (2003)

187. Lu, Y., Cohen, I., Zhou, X.S., Tian, Q.: Feature selection using principal feature analysis. In: Proceedings of the 15th International Conference on Multimedia, pp. 301–304 (2007)

Printed in the United States
By Bookmasters